Baboon Mothers
and Infants

Baboon Mothers and Infants

Jeanne Altmann

HARVARD UNIVERSITY PRESS
Cambridge, Massachusetts
and London, England 1980

Library of Congress Cataloging in Publication Data

Altmann, Jeanne.
  Baboon mothers and infants.

  Bibliography: p.
  Includes index.
   1. Yellow baboon—Behavior.   2. Parental behavior
in animals.   3. Animals, Infancy of.   4. Mammals—
Behavior.   5. Mammals—Kenya—Amboseli Game Reserve.
I. Title.
QL737.P93A39   599'.884   79-21568
ISBN 0-674-05856-9

To
my parents, Lillian and Oscar Glaser,
my husband, Stuart,
my children, Rachel and Michael,
simultaneously three generations and one

Acknowledgments

THIS STUDY HAS SPANNED two continents over a number of years, benefiting from the contributions of many people. It is part of the overall research on baboon ecology and behavior initiated by Stuart Altmann. The government and people of Kenya have provided the research access and cooperation that have made our projects possible. During the mother-infant study, Joseph Kioko as warden of the Amboseli National Park was supportive and encouraging, as Daniel Sindyo had been in previous years and the current warden, Bob Oguya, is now. Christina Kioko's early tabulations of data contributed greatly to the shape of the eventual analyses. Through the years our camp worker, Mwongi Kirega, has freed many hours for field work that would not otherwise have been available.

Financial support has been provided by the American people through grant MH 19617 from the U.S. Public Health Service. In Chicago, Rebecca McCauley has seen the data through every step of analysis from the first conversion into computer files through the preparations of the figures in this text. Her thoroughness and flexibility have been invaluable. In the last few months she has been joined by Maureen Condic, who has helped with preparation of tables and figures, and by Joan Luft, who assisted with preparation of the index and has checked and retyped more drafts of this manuscript than I can remember. Steven Muka's and Kenneth Kaye's development of the CRESCAT computer system and their assistance as I began to use it facilitated data analysis. Stevan Arnold, Donald Fiske, Daniel Freedman, Robert Hinde, Kenneth Kaye, Martha McClintock, David Post, Richard Shweder, David Stein, Michael Wade, Jeffrey Walters, and Anne Wells provided valuable comments on an earlier draft of the manuscript.

At every stage of the mother-infant study I have benefited from the perceptive and wide-ranging questions, comments, and critiques of Mihaly Csikszentmihalyi. Our friend and colleague Glenn Hausfater has shared with Stuart Altmann and me the responsibility and work involved in the Amboseli longitudinal project, has shared ideas, good times and difficult ones, and provided valuable comments on this manuscript.

Stuart Altmann has contributed immeasurably to every aspect of my education, my research, and my life. Our field work has always been a family enterprise: our children, Michael and Rachel, have patiently and enthusiastically shared our life in the field, sometimes assisted with the research in Kenya and in Chicago, and tolerated my frequently preoccupied air during the recent months of writing.

Contents

Figures

Tables

1 / Introduction

ONE OF THE FIRST things that strikes anyone who follows a group of baboons across the African savannahs is how much time and energy they spend just making a living: feeding, walking, avoiding predation. The surprise is particularly great for those who have watched captive monkeys, or even have seen films of wild primates, for these films are inevitably constructed from the 10 percent (or less) of the daytime that the animals spend socializing, not the 65 to 70 percent of the time they spend walking and feeding (Post 1978, Rose 1977, Slatkin 1975). One can easily overlook these activities of maintenance or production. In captivity such activities occupy only 10 to 20 percent of the day even in large enclosures or small islands with provisioning (see, e.g. Fisler 1967, Post and Baulu 1978, Rasmussen and Rasmussen 1979) and typically involve nothing more than picking prepared food pellets out of a hopper. In the baboons' natural habitat, however, it is impossible to ignore these activities when animals walk several kilometers a day in the hot sun (S. Altmann and J. Altmann 1970), when they laboriously dig bulbs or grass corms from the ground for much of the day, and when their rest periods seem necessary for recovery from fatigue, rather than just a response to boredom. Thus, in our studies of yellow baboons, *Papio cynocephalus,* in Amboseli National Park, Kenya, my colleagues and I have been led—no, virtually forced—by our animals to consider their lives as an integrated whole and not as composed of independent social and nonsocial pieces. In all likelihood, the fact that they spend three-quarters of their day making a living in itself affects their other activities; and not just in how much time they have available for those other activities, which is immediately obvious, but in much more complex ways as well. It is some of these complexities that I shall explore in the following analyses.

1

Fig. 1. *A group of yellow baboons feeding with Kilimanjaro in the background; Amboseli National Park, Kenya.*

Two particularly crucial life stages for any primate are those of motherhood and infancy. Most field studies of wild primates (e.g., DeVore 1963, Jay 1963, Lindburg 1971, Poirier 1968, Ransom and Rowell 1972, Van Lawick-Goodall 1967, 1971) have included qualitative descriptions of mothers and infants, and some provide quantitative results for a few behaviors or for a few typical individuals. In addition, quantitative studies of captive nonhuman primate mothers and infants have been conducted on a variety of species and have included a wide range of social settings, from isolation studies to those conducted with small groups of mixed ages and both sexes. Most of these studies have focused on infants rather than mothers and have emphasized predictions of adult behavior and effects of mother-infant separation as part

Fig. 2. *Amboseli baboons (Alto's Group) resting and grooming.*

of an attempt to provide a monkey model for human mother-infant separation. They have included both "purely" observational and more experimental investigations. However, with but rare exceptions (e.g., Draper 1976 and Konner 1976 for humans, Struhsaker 1971 for vervet monkeys), systematic quantitative studies of mothers and infants in the complexity of natural settings have been absent from research on both human and nonhuman primates, and studies of environmental influences have been limited primarily to a few elements in the social environment.

This study is an attempt to assess the nature and extent of external influences on baboon mothers and their young infants. My general goal was to identify and measure factors affecting survival and behavior during motherhood and infancy, and to identify likely ontogenetic origins of differences in adult behavior and life history patterns. Toward this end I collected systematic ecological, demographic, and behavioral data on all baboon mothers in a group in Amboseli National Park, Kenya, for which long-term life history data were available on individual members.

The project had several specific goals generated by the overall aims. First, I sought an understanding of the demography of motherhood and infancy. Are these high-risk periods for baboons, as they are for other animals? To answer this, I turned to the demographic data

Fig. 3. *Juvenile female Fanny* (left) *sits near her mother, Oval, who is nursing nine-month-old Ozzie, and is watching as one adult female grooms another who has a very young, black infant.*

that my colleagues and I collected from 1971 through 1978. Such information accumulates slowly, and David Post, Jane Scott, Sue Ann McCuskey, Francis Saigilo, Jeffrey Walters, David Stein, and especially Stuart Altmann and Glenn Hausfater have all contributed to its collection. From these data I was also able to examine the effects of maternal age, dominance rank, and infant gender on mortality and the effects of maternal dominance rank on infant gender (Chapter 4).

Next, I sought a quantitative description of the nonsocial milieu or ecology as it affected mothers and infants (Chapter 5) and a test of the extent to which this ecology limits infant survival and the timing of births and of weaning. Further, I asked whether mothers' time budgets were affected by their dominance rank and, if so, whether these effects were reflected in differential survival or in the time or attention available for infant care. These topics are treated in Chapter 5. The effects of mothers' time budgets on aspects of mother-infant relationships are then considered in Chapters 8 and 9.

Another set of related questions involved the social milieu. In what ways does a female's social life change when she gives birth? Which changes are beneficial and which harmful? What are the nature and sources of individual differences in social milieu among mothers?

In the literature of behavioral biology, much attention has been directed to the relationship between male dominance rank and reproductive success (see Bernstein 1976, Hausfater 1975a, Packer 1979b, and references therein); yet for a number of primates female dominance ranks have been shown to be more stable both within and between generations (Hausfater 1975a, Kawai 1958, Kawamura 1958, Missakian 1972, Sade 1967). Few field studies identify either the ontogenetic origins of female dominance relationships (Cheney 1977 for chacma baboons; Kawai 1958, Kawamura 1958 for Japanese macaques) or their potential evolutionary consequences (Dittus 1977 for toque monkeys; Drickamer 1974 for free-ranging, provisioned rhesus monkeys; Dunbar in preparation for geladas). Can we identify the origins of dominance rank "inheritance" in these early social interactions? Who are the individuals who constitute the social world of mothers and infants? Are they members of particular age classes? Are they particular individuals? Are the individual preferences in choice of associates observed in sexual consortships (Hausfater 1975a, Packer 1979b) also manifested at this time? These are some of the questions addressed in Chapter 6.

With an understanding of the ecological and social milieu and the demographic (e.g., age, parity) and sociological (e.g., dominance rank) characteristics of the mothers, I could determine not only whether these variables affect each other, affect mothers directly, and affect infant mortality, but also whether they affect maternal care, the mother-infant relationship, and infant development. Thus in Chapters 7 and 8 I provide a quantitative description of maternal care and infant development and examine the sources and consequences of the individual differences that emerge.

Recently, models of the evolution of parent-offspring relations (Alexander 1974, Blick 1977, Charlesworth 1978, MacNair and Parker 1978, Parker and MacNair 1978, Stamps et al. 1978, Trivers 1974) have been developed to explain phenomena such as weaning conflict and to make predictions about aspects of parental care that evolve through natural selection. Models of the evolution of behavior are economic models; their major parameters include costs and benefits of behavior, usually measured in the "currency" of fitness, of survival and reproduction, and of relative representation of genes in the next generation. Assuming a degree of heritability of the behavior in question, one is then concerned with attempts to estimate these costs and benefits. Thus, one of the questions I shall address in subsequent chapters is the costs and benefits of each behavior I discuss. Does the behavior appear to be selectively advantageous or disadvantageous? The answer, of course, will not tell us whether any behavior evolved through natural selection. However, given certain assumptions about the population, the mating system, and the heritability of the traits involved, the answer will suggest whether the behavior could have spread at least partially through natural selection. In Chapter 9 I consider these models from the holistic perspective of the current study and indicate how such a perspective and the available evidence for baboons suggest additional or alternative views and needed research.

This book is about motherhood and infancy more than about "mothers" and "infants": motherhood and infancy are only two stages, albeit particularly important ones, in the life histories of individuals whose lives extend before and, one hopes, after these periods. Females enter motherhood with their pasts and they and their infants carry into their futures the marks of their experiences during the life stage they so intimately share. Even during this period females' lives remain more complex and multifaceted than the label "mother" or "motherhood" implies. It is the pasts and the futures and the concurrent facets of their lives that I shall examine and relate to the experiences of motherhood and infancy.

In all aspects of the present study, one fact recurs: baboon mothers, like most primate mothers, including humans, are dual-career mothers in a complex ecological and social setting. They do not take care of their infants while isolated in small houses or cages as the rest of baboon life goes on. They are an integral part of that life and must continue to function within it. The baboon world affects them, and they it, throughout their lifetime. Determining the consequences of limited available time and the effects of maternity, social milieu, and physical environment on individuals' time budgets constitutes a major thrust of this investigation.

For short-lived animals the life history approach to the study of behavior is common. In studies of primates, including humans, it is much rarer, partially because life histories are so long and each stage is sufficiently complex to keep researchers themselves occupied for whole lifetimes. The work that follows is one, necessarily incomplete, attempt to combine a life history approach with the quantitative study of primate behavior in the field.

The general problems that face most human and nonhuman primate mothers are in many ways similar. High rates of infant mortality and appreciable maternal risk have probably been characteristic of most human and other primate populations. Rates of birth and death will determine the relative ages of siblings, affect the size of age cohorts and other aspects of the social group in which mothers and infants find themselves (S. Altmann and J. Altmann 1979). Likewise, ecological and social factors that affect mortality will be important aspects of motherhood and infancy.

The complex ways in which members of society provide support and also transmit existing social structure to infants are important for any primate mother. If we can begin to understand the origins of normal differences in maternal behavior, we shall be better able to predict the consequences of changes in these factors.

By examining the problems of motherhood and infancy in detail for one primate species we may learn questions to ask, parameters that should be measured, and possible strategies for studying complex ecological and social relationships in related species. Just as the study of another human culture sometimes provides a different perspective, new insights, occasionally new answers, and more often new questions, I hope that this attempt to deal with the complexity of experience in a study of baboon motherhood and infancy will strike some familiar chords or stimulate some new ideas in those whose main interest is in another species, perhaps even our own.

2 / Baboons and Their Habitat

Baboons and Behavioral Research

BABOONS (*Papio* spp.) are large, group-living, Old World monkeys, full-grown males weighing about 23 kilograms, females 11 to 12 kilograms (calculated from data in Bramblett 1969). There are several species of baboons, exhibiting various social structures and occupying a wide variety of African habitats. Baboons are among the most terrestrial of the monkeys, a feature that makes them more observable than most other primates and has led also to attempts to regard them as models for hominid evolution (e.g., DeVore and Washburn 1963, Jolly 1970).

Despite the widespread use of baboons in medical research, few behavioral studies have been conducted on members of the genus in captivity. Major early exceptions were Hans Kummer's study of hamadryas baboon (*Papio hamadryas*) social organization in the Basel Zoo (Kummer and Kurt 1965) and Thelma Rowell's investigations of captive olive baboons (*P. anubis*), particularly of female social relations and infant development (Rowell 1966a, 1968, 1969, Rowell et al. 1968). Terrence Anthoney (1968) and Gilbert Boese (1975) studied the Brookfield, Illinois, colony of guinea baboons (*P. papio*), and more recently Anthony Coelho, Claud Bramblett and colleagues (e.g., Young and Bramblett 1977, Young and Hankins 1979) have begun long-term studies of several *Papio* species and hybrids at the Southwest Foundation for Research and Education. In contrast, rhesus monkeys (*Macaca mulatta*), other macaque species, and squirrel monkeys (*Saimiri sciureus*) have been major subjects of behavioral investigations for many years, especially at the Japanaese Primate Centre and in laboratories such as those of Bernstein, Harlow, Hinde, Jensen, Kaufman,

Mason, Mitchell, Ploog, Rosenblum, and Sackett, where research has often focused on infant development and the ontogeny of behavior.

Field studies of the various *Papio* species have had a very different history. At the turn of the century Eugene Marais lived for several years with chacma baboons (*P. ursinus*) in South Africa. Only recently, many years after his death, was his unfinished manuscript found and published (Marais 1969), providing a sometimes strange but intriguing view of baboon life. In the 1950s field studies began in earnest; research on chacma baboons by Bolwig and by Hall and on olive baboons by DeVore were the first of what have become a growing variety of investigations of the several baboon species. Kummer's field study of social organization of hamadryas baboons, Stolz and Saayman's of the behavior of chacma baboons, Thelma Rowell's field research on olive baboons, and the Altmanns' study of the ecology of yellow baboons all followed shortly after, in the 1960s. Most of these studies were general investigations producing qualitative or quantitative descriptions of aspects of ecology or behavior. An annotated bibliography of the early field studies has been published by Baldwin and Teleki (1972). The less closely related gelada baboon, *Theropithecus gelada*, has been the subject of research for a number of years in the Semien mountains of Ethiopia, particularly by Crook, the Dunbars, Kawai, Osawa, and Mori.

The 1970s have seen two major changes in the conduct of baboon field studies. First, research is more focused and problem-oriented and is often concerned with hypothesis testing: following up on questions raised by the more general studies, by theoretical considerations (e.g., foraging and kin selection theory), or by laboratory investigations of other species. This change was foreshadowed in the second wave of studies in the 1960s and is partially paralleled by changes in field studies of other species. Second, long-term studies were initiated at several sites. These two changes, combined, are beginning to result in longitudinal, life history studies of known individuals and in-depth investigations that have involved researchers from a wide variety of disciplines. The sites with appreciable continuity include those of Kummer, Abbeglen, and colleagues with hamadryas baboons in Ethiopia; Ransom, Packer, and others with the olive baboons at Gombe in Tanzania; the Altmanns, Hausfater, and colleagues with the yellow baboons in Amboseli, Kenya; Harding, Strum, and others in Gilgil, and more recently Popp and DeVore at Mara, all with olive baboons in Kenya; W. J. Hamilton with chacmas in Botswana; and the Rasmussens and Rhine with yellow baboons in Mikumi, Tanzania.

General Natural History

Savannah baboons are born into semiclosed groups of about 40 animals, but ranging from less than 10 to almost 200 (see review in S. Altmann and J. Altmann 1970). In the wild, baboon females reach menarche at four-and-a-half to five years of age (J. Altmann et al. 1977, Packer 1979a, Sigg and Kummer in preparation), and then before their first conception they experience a series of menstrual cycles for about a year ("adolescent sterility") (J. Altmann et al. 1977). Their infants are born after a gestation period of six months (177 days). Postpartum amenorrhea follows, such that successive infants are produced at intervals of one-and-a-half to two years unless the process is accelerated by the death of the previous infant (J. Altmann et al. 1977, 1978). Thus, after a female reaches full maturity, that is, experiences her first pregnancy, she spends approximately half her life with a dependent infant, almost a third of her life pregnant, and the remaining time undergoing·menstrual cycling. This is true throughout adulthood; menopause has not been demonstrated for any nonhuman primate. Fig. 4 depicts the reproductive stages experienced by adult females during a two-and-a-half-year period. Several females reached menarche and gave birth to their first infants during this period, several died, one experienced menstrual cycles the whole time without becoming pregnant, others sequentially experienced several different reproductive stages.

The Amboseli Population

The cynocephalus, or yellow, baboons include those we studied in the short-grass savannah region of southern Kenya just a few miles north of Mt. Kilimanjaro. There they inhabit Amboseli National Park along with a wide variety of other animals (S. Altmann and J. Altmann 1970, Struhsaker 1967, Western and van Praet 1973). Some of these, such as leopards, lions, and most recently hyenas (J. Stelzner in preparation), prey on baboons (J. Altmann 1978, S. Altmann and J. Altmann 1970, DeVore and Hall 1965) (Fig. 5). The baboons themselves prey on others. These prey are usually grasshoppers and other invertebrates, but, they also include African hares (Fig. 6), gazelles, and vervet monkeys (see, e.g., S. Altmann and J. Altmann 1970, Harding 1973, Rowell 1966b, Strum 1975), especially during the dry season (Hausfater 1976). Most other large mammals in the area, such as elephants, wildebeest, and zebras, have an essentially neutral relationship with baboons, or an indirect one, through plants (Fig. 7).

The Amboseli baboons have two sources of contact with humans

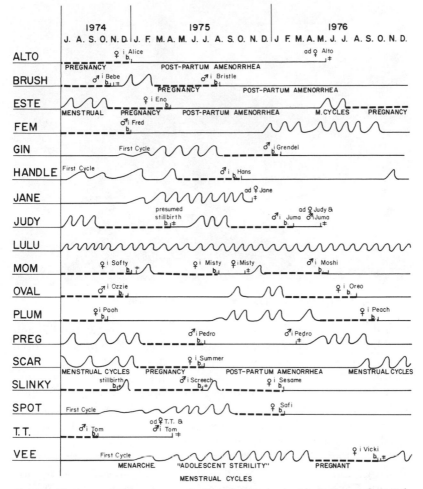

Fig. 4.   *The reproductive stages of all adult females in Alto's Group from July 1974 through December 1976. On the line for each female infant births are indicated by (b), deaths by (≠). Menstrual cycles are indicated by a wave roughly corresponding to the change in sex skin swelling during each cycle. Before each birth is a period of pregnancy (indicated by a dotted line), during which swelling ceases, and following each birth a period of postpartum amenorrhea. For several females (Alto, Brush, Este Scar, and Vee), these stages are labeled below the line. Note that Gin, Handle, Spot, and Vee all reached menarche—experienced their first menstrual cycles—during this period, and that Lulu cycled continuously without becoming pregnant.*

Fig. 5. *A lion eating a subadult or adult male baboon.*

other than the observers. First, Maasai pastoralists live within the ani-mals' home range during the dry seasons. Although the Maasai seldom overtly interfere with most wildlife, their cattle constitute a large pro-portion of the Amboseli biomass during the dry seasons (Western 1973) and compete with other grazers. Baboons, along with other wildlife, are displaced by cattle herds. Additionally, Maasai dogs and children occasionally chase baboons. Since late 1976, the Maasai have been excluded from Amboseli National Park in exchange for ser-vices and alternative resources, and have been provided with financial compensation for allowing wildlife to graze on Maasai land outside the park. Contact between baboon groups and Maasai and the effect of competition with Maasai cattle are now restricted to portions of the baboons' home range that are outside the park.

Second, tourists visit Amboseli. In recent years the number of tourists has greatly increased (Western 1977); but in contrast to the tourists of the early 1960s, these visitors are usually non-Kenyans and often on commercial tours. The result is that they spend little time in the park, almost all of it within a small area near the lodges (Western 1977), several miles from our main study area. Thus, members of our main study group currently seem to come in contact with tourists even less than study groups did when we first worked in Amboseli in the early 1960s.

Fig. 6. *Female Scar eating a hare.*

Fig. 7. *Baboons in Amboseli feeding in the open acacia woodland among zebras and impala gazelles.*

Western and van Praet (1973) have described the recent transformation of the Amboseli habitat from an acacia woodland to a dry habitat with salt-loving plants. Several years of heavy rainfall during the 1960s raised the water table to unusually high levels. The rising water led to the death of most of the trees in the acacia woodland. Some trees died as a result of salts, which, through "wicking action," form a layer in the soil just above the water table and are carried up into the root zone of the trees. The roots of other trees were completely drowned (Western and van Praet 1973). Overgrazing by large herds of Maasai livestock reduced available grazing material, and this depletion, in turn, probably resulted in a shift in diet of animals that are both browsers and grazers, most notably elephants, and led to further damage to trees. Overgrazing was aggravated by poor rainfall starting with a drought in 1969.

As a result, major declines occurred in populations of many species of wildlife that depend on woodland habitats (Western and van Praet 1973), including the vervet monkeys (Struhsaker 1973, 1976). Over two thousand baboons were censused in the Amboseli waterhole area during 1963–64 (S. Altmann and J. Altmann 1970), but by 1971 the baboon population in the same area was only about 10 percent that size (Hausfater 1975a). The percentage decline in the baboon population was considerably greater than that of the vervets (90 percent versus 33 percent), at least partially owing to the trapping of approximately 230 baboons in 1965 by the Southwest Foundation for Research and Education (W. Maples, personal communication); also probably partially owing to the presence of a paralytic virus in the population in 1964 (S. Altmann and J. Altmann 1970).

In 1971 we began a general program of periodic monitoring of the baboon population as a whole and longitudinal monitoring of demographic events in a single study group. A detailed analysis of the baboon population since 1971 is under way (J. Altmann et al. in preparation) but the most relevant demographic data for the subjects of the longitudinal study, Alto's Group, are described here and in Chapter 4.

Alto's Group

The subjects of this study and of most of our detailed longitudinal observations are members of Alto's Group, one of five groups now occupying the central waterhole area of Amboseli. Longitudinal data on individuals in Alto's Group have been collected since 1971 by a succession of field workers (e.g., J. Altmann 1978; J. Altmann et al. 1977, 1978; Hausfater 1975a, 1976; Post 1978; Slatkin 1975; Slatkin and Hausfater 1976). In the fall of 1972, a fusion was completed between the original Alto's Group and a one-male group, High Tail's Group (J. Altmann et al. 1977, McCuskey 1975).

Of the 35 members of Alto's Group in July 1971 and the 15 members of High Tail's Group when they were first censused in November 1971, 23 from Alto's Group and 8 from High Tail's Group remained in the combined Alto's Group of 47 members four years later, at the onset of this study in July of 1975. Thus there was approximately a 40 percent turnover of group members from 1971 to 1975. This turnover was due to births, deaths, and migrations. Of course, even more such demographic events occurred during the intervening years than are indicated by these figures, since many individuals were born and died, joined the group and then either died or emigrated, and so on. Consider Dogo, an infant born to female Preg in August 1971 (see Appendix 1). He has grown up in an ever-changing social group and has known approximately 68 animals, while staying in his natal group of about 45 animals. None of his cohort from 1971–72 remained in Alto's Group in 1975; yet almost all of his younger sister's 1973 cohort did. Most of those who were juveniles when he was born were still in the group in 1975, though some males had temporarily emigrated and then returned. Of the eight males who were fully adult when Dogo was born, only three, Peter, Max, and Stubby, remained. In January 1979, Dogo was just reaching adulthood when he was killed by two hyenas (J. Stelzner in preparation). At the time of his death, he was the oldest animal that we knew from birth. His mother and two younger sisters remained in the group; a younger brother, Pedro, one of the subjects of the current mother-infant study, died when eight months old.

In the results of this study I document a more complicated picture of motherhood and infancy than has usually been considered. Yet I still deal primarily with only one year and one reproductive stage in the lives of adult females: their last month of a pregnancy and the first year of their infants' lives. Occasionally I shall hint at the longer view: relationships established before infants' birth, bonds that persist for infants as they mature, lifetime dominance relations, and long-term ecological patterns. Many of these are currently topics of Amboseli research conducted by Stuart Altmann, Glenn Hausfater, David Post, David Stein, and Jeffrey Walters. However, the continuities and the long-term bonds will also be set against the ever-changing demographic panorama outlined above.

Table 1 provides a list of the members of Alto's Group at the onset of this study, and a list of demographic events during the period from 1 July 1975 through 31 October 1976 is provided in Table 2. The frequent male migrations that occurred during this year are a reflection of the maturation of several subadult males and the resultant fighting and changes in dominance rank that this precipitated. These rank changes, mortality, and emigrations affected all group members but had especially strong effects on individual mothers and infants with whom the relevant males were associated. Maternal genealogies are drawn in Appendix 1. Appendix 2 contains case history descriptions of the mothers and infants. Brief descriptions of the adult males are in Appendix 3. The reader is referred to these appendices for more general individual descriptions and impressions of the major cast of characters who generously provided the data points for the chapters that follow.

Each year approximately ten infants are born into the group of

Table 1. Membership in Alto's Group at the beginning of the study, 1 July 1975. [a]

Adult males	Subadult-young adult males	Juvenile males	Adult females (infants) [b]	Juvenile females
Slim (SM)	Stiff (SI)	Swat (SW)	Alto (AL) (♀ Alice) (AC)	Janet (JT)
Stubby (SB)	Red (RE)	Nog (NO)	Spot (SP)	Dotty (DT)
Even (EV)	Russ (RS)	Toto (TO) [c]	Mom (MO)	Striper (ST)
Ben (BN)	Stu (SU)	Dogo (DG)	Lulu (LU)	Nazu (NZ)
Peter (PE)	Total: 4	Jake (JK)	Vee (VE)	Cete (CT)
Max (MX)		Total: 5	Preg (PR)	Fanny (FN)
High Tail (HT) [c]			Scar (SC)	Total: 6
B.J. (BJ)			Oval (OV) (♂ Ozzie) (OZ)	
Chip (CP)			Fem (FM) (♂ Fred) (FR)	
Dutch (DC)			Gin (GN)	
Total: 10			Jane (JA) [c]	
			Slinky (SK) [c]	
			Handle (HA) [c]	
			Plum (PL) [c] (♀ Pooh) (PH)	
			Brush (BR) [c]	
			Este (ES) [c] (♀ Eno) (EN)	
			Judy (JD)	
			Total: 17 (+ 5 infants)	

a. Animals are ordered by dominance rank within each class. Two-letter abbreviations follow each name. These will be used in subsequent tables and figures where space does not permit use of full names.

b. The first year of life includes the whole period of strong dependence on the mother, and it is this first year to which I refer when I use the term "infant." However, no major physical change occurs at a year of age. Nor does the period of dependency end sharply or completely at that age, a fact that has probably led to errors of age estimation in studies of individuals who are first encountered after the early months of life (J. Altmann et al. 1977).

c. Members of High Tail's Group at the time the groups fused in late 1972.

Table 2. Demographic changes in Alto's Group,
1 Juiy 1975–31 October 1976.

Date	Change	Animal	New group size
7 July	Birth	♀ Summer (SR) (of Scar)	48
9 July	Birth	♂ Pedro (PD) (of Preg)	49
15 Aug.	Birth	♀ Misty (MS) (of Mom)	50
17 Aug.	Birth	♂ Bristle (BS) (of Brush)	51
18 Aug.	Emigration	Ad ♂ Dutch	50
20 Aug.	Immigration	Ad ♂ Dutch	51
29 Aug.	Death	Ad ♂ Dutch	50
15 Oct.	Birth	♂ Hans (HN) (of Handle)	51
23 Oct.	Death	Ad ♀ Jane	50
25 Oct.	Death	i ♀ Misty (of Mom)	49
13 Nov.	Death	Ad ♂ Stubby	48
27 Jan.	Birth	♂ Grendel (GD) (of Gin)	49
2 Feb.	Birth	♀ Sesame (SS) (of Slinky)	50
5 Feb.	Emigration	Ad ♂ Russ	49
5 Feb.	Emigration	Ad ♂ Stiff	48
7 Feb.	Immigration	Ad ♂ Russ	49
9 Feb.	Immigration	Ad ♂ Stiff	50
12 Feb.	Emigration	Ad ♂ Stiff	49
28 Feb.	Birth	♂ Juma (JM) (of Judy)	50
2 Mar.	Birth	♀ Safi (SF) (of Spot)	51
7 Mar.	Emigration?	Ad ♂ B.J.	50
7 Mar.	Emigration?	Ad ♂ Chip	49
7 Mar.	Emigration	Ad ♂ Stu	48
8 Mar.	Immigration	Ad ♂ Stu	49
12 Mar.	Death	i ♂ Pedro	48
22 Mar.	Emigration	Ad ♂ Russ	47
24 Mar.	Immigration	Ad ♂ Russ	48
26 Mar.	Emigration	Ad ♂ Russ	47
26 Mar.	Immigration	Ad ♂ Russ	48
16 Apr.	Emigration	Ad ♂ Red	47
16 Apr.	Emigration	Ad ♂ Russ	46
3 May	Emigration	Subad ♂ Swat	45
10 May	Death	Ad ♀ Judy	44
10 May	Death	i ♂ Juma	43
16 May	Death	Ad ♂ High Tail (leopard)	42
21 May	Death	Ad ♀ Alto	41
24 May	Emigration	Ad ♂ Even	40
25 May	Immigration	Ad ♂ Even	41
28 May	Emigration	Ad ♂ Even	40
28 May	Immigration	Ad ♂ Red	41

Table 2 (*continued*).

Date	Change	Animal	New group size
29 May	Emigration	Ad ♂ Red	40
29 May	Immigration	Ad ♂ Even	41
30 May	Immigration	Ad ♂ Red	42
1 June	Birth	♂ Moshi (MH) (of Mom)	43
9 June	Emigration	Ad ♂ Red	42
12 June	Immigration	Ad ♂ Red	43
20 June	Emigration	Ad ♂ Red	42
21 June	Immigration	Ad ♂ Red	43
22 June	Emigration	Ad ♂ Red	42
23 June	Immigration	Ad ♂ Red	43
30 July	Birth	♀ Oreo (OR) (of Oval)	44
22 Aug.	Death	Subad ♂ Nog	43
30 Aug.	Immigration	Ad ♂ Lip	44
1–2 Sept.	Emigration	Ad ♂ Lip	43
4 Oct.	Birth	♀ Vicki (VK) (of Vee)	44
5 Oct.	Emigration	Ad ♂ Stu	43
6 Oct.	Immigration	Ad ♂ Stu	44
8 Oct.	Emigration	Ad ♂ Stu	43
18 Oct.	Emigration[a]	Juv ♀ Pooh	42
19 Oct.	Death[b]	Juv ♀ Pooh	42
19 Oct.	Immigration	Ad ♂ Stu	43
20 Oct.	Emigration	Ad ♂ Stu	42
23 Oct.	Emigration	Ad ♂ Red	41
24 Oct.	Immigration	Ad ♂ Red	42
27 Oct.	Death	i ♀ Vicki	41
28 Oct.	Emigration	Ad ♂ Red	40
29 Oct.	Immigration	Ad ♂ Red	41
31 Oct.	Emigration	Ad ♂ Red	40

a. Abandoned by group owing to severe wounds suffered 17 Oct.
b. Group joined Pooh, but again abandoned her; Pooh not seen again.

about 45 animals. Usually six or seven infants survive the first year of life. In recent years, Alto's Group has remained at a fairly stable population level, and on the basis of age-specific birth and death rates, we predict that this stability will continue (J. Altmann et al. in preparation). These were essentially the demographic conditions during the study of mothers and their newborn infants that I conducted from July 1975 through July 1976, during October 1976, and during a short pilot study in 1974.

The Mother-Infant Study

All females in Alto's Group who had infants under a year of age were included in the study. Except for one adult female, Lulu, who has not conceived since at least 1971, probably not since 1969, all adult females in Alto's Group had at least one infant during this study, including four primiparous females and two females who experienced their first viable delivery during this period. Infant maturation and mortality factors resulted in three females being included with two successive infants. Five infants who were included were born before July 1975. Thirteen others were born during the study.

When I arrived in Amboseli at the end of June 1975, several infants were already in the group. Information about them was kindly provided by David Post and Jane Scott. The oldest was Pooh, a scrawny eight-month-old female with locomotor impairment who was Plum's first-born. Several other infants were born about the same time as Pooh and would have been included in the study if they had survived. Slinky's first infant was stillborn in November. Brush's first infant was born in October with congenital deformities; it lived only a few weeks. A high-ranking older female, T.T., had given birth to a healthy male infant, Tom, within a week of Pooh's birth. I was quite disappointed when we learned of T.T.'s and Tom's sudden disappearance in May.

Four females gave birth within two weeks at the end of 1974, when Pooh was about two months old. Alice was born to high-ranking, elderly Alto, who had two other known surviving daughters in the group, Spot and Dotty. Another high-ranking older female, Mom, also gave birth to a daughter, Softy at that time, but Softy died of unknown causes after only a few weeks. Mid-ranking Oval and Fem both gave birth to sons, Ozzie and Fred. Oval's juvenile daughter, Fanny, was in the group. Being Fem's first infant, Fred had only one "known" relative, his mother's putative sister, Gin.

In addition to Pooh, Alice, Ozzie, and Fred, two-month-old Eno was already in the group when I arrived. Early one evening in late April she was born to Este, an elderly low-ranking female who had a five-year-old son, Toto, in the group when Eno was born. We had expected Eno to have a close peer, because Judy, another elderly low-ranking female, was due to give birth in March. However, Judy temporarily disappeared from the group at about that time, and was wounded and apparently experienced a stillbirth before her return. At the end of June Slinky gave birth to her second infant. Although it apparently was well formed externally, it was rigid at birth and died two days later.

In July Pooh, Alice, Ozzie, Fred, and Eno were joined by Summer and Pedro, offspring of mid-ranking Scar and Preg, who already had surviving offspring in the group. In August, Misty and Bristle were born within a few days of each other. Misty was the daughter of high-ranking Mom. Bristle was Brush's son, and since Brush's first infant had died so early and had been malformed, it was almost as if Bristle was her first-born. The fact that Brush was very low-ranking made a striking contrast in the experiences of these two infants, one that had not been so evident with the more closely matched Summer and Pedro. Misty's death just after she turned two months old was a particularly unfortunate one.

Just before Misty's death, Handle, a low-ranking female, gave birth to her first infant, Hans. Hans's environment was quite similar to that of Bristle. Since their mothers were also rather frequent associates and since no other infants were born until late January, these two infants spent much time together.

Starting in late January, four infants were born within a month of each other. Gin gave birth to her first infant, Grendel. A few days later, Slinky finally gave birth to a viable infant, although Sesame appeared weak from birth and had bouts of illness periodically thereafter. At the end of February, Judy again gave birth, this time to a healthy son, Juma. Judy's daughter, Janet, was almost three years old by then. A few days later, Alto's daughter, Spot, also high-ranking, gave birth to female Safi. With two sisters and a grandmother in the group, Safi had more known relatives than any of the other infants.

Several months passed. One night during this period Judy and Juma disappeared and did not return (see Appendix 2). Then at the beginning of June, Mom gave birth to Moshi. I was struck with how similar Moshi's world was to that of his sister, Misty, and how different it was from that of infants such as Bristle who had been born at the same time as Misty. Similar contrasts were evident among the other infants. I shall explore the nature and origins of some of these differences in the chapters that follow. Often Bristle and Moshi will provide useful examples of constrast.

Just after I left Amboseli at the end of July, Ozzie's younger sister, Oreo, was born. When I returned to Amboseli for a month that October I was able to observe Oreo as well as Vicki and Peach, who were born at the beginning of October. Despite over a year of observations on new mothers, including the first day of life for most of the infants, I had not actually been present for a birth. It was a particularly exciting surprise when the last birth, that of Plum's infant, Peach, occurred while I was making observations one afternoon, and when Plum, the most wary female in the group, tolerated my presence at a moderate distance, enabling me to watch most of the labor and birth.

3 / Methods

BECAUSE MOTHERS WITH NEONATES appear to be the class of individuals in our baboon population that are most sensitive to being observed, and because most of our previous research had been done from atop a vehicle (S. Altmann and J. Altmann 1970), the first month of this study and to some extent the second were partially devoted to accommodating the females to observation on foot within 5 to 10 meters (Fig. 8). Much closer distances were possible for most mothers, but consistent with the aim to minimize our effect on the system we were studying, I always tried to stay at a distance that would not discourage the mothers' interactions with the most sensitive individuals in the group. In addition, some types of detailed data were considered sufficiently reliable to use only after several weeks or more of systematic sampling. Consequently, for some analyses I used data obtained from July 1975 onward; for others, only data obtained later.

After various sampling schemes were tried, the following scheme was established by September 1975. Each female was sampled during the last month of pregnancy for two days at least a week apart, on the day of birth and the fifth day of infant life, on two days during the infant's second week, and one day per week thereafter until the infant was six months old. During the next six months samples were taken two or three days per month. Whenever impassable mud, illness, or other factors reduced available observation days, sampling mothers with older infants was sacrificed in favor of sampling those with younger ones.

In gathering behavioral data I sought samples that would provide unbiased estimates of rates of behavior, of time budgets and bout durations, and other measures that are derived from these. Focal-animal

Fig. 8. *The author collecting data, indicating the observation distance and relative neutrality with the animals. Gradually through the years we have been able to achieve a situation in which animals rarely direct any behavior toward us: not fear, as occurs in the early years of study, nor aggression or affiliative behaviors, which are likely to occur after animals lose their fear. Maintaining this condition requires an active effort on our part but the success has been more than sufficient reward.*

(continuous) sampling and instantaneous (point) sampling are the most versatile and suitable of the common techniques used to provide unbiased estimates of time budgets; focal samples provide the data needed to estimate rates and bout durations (see J. Altmann 1974, Kraemer 1979, Simpson and Simpson 1977). Unfortunately, until recently most field studies have utilized ad libitum sampling and many laboratory studies one-zero sampling (J. Altmann 1974), neither of which provides unbiased estimates of these parameters. Detailed comparisons of bouts, rates, time budgets, and derived measures within and between studies that utilize these techniques cannot be made unless the sampling scheme itself is considered as a possible source of the differences and similarities found in the results (J. Altmann 1974, Kraemer 1979). Consequently, I shall primarily restrict detailed comparisons between the results of this study and others to those comparisons not likely to be confounded by sampling biases. More generally, I shall emphasize, where possible, comparisons with those studies that differ in the fewest major variables, such as species and extremes of caging and rearing conditions, so that similarities or differences can be at least provisionally attributed to characteristics of the animals in their normal range of living conditions.

On each sample day, I sampled the behavior of two females. They were sampled alternately, each for 15 minutes in-sight time (or 20 minutes in the first months of the study) out of every hour from 0800 through the 1700 hour but excluding the 1200 hour. The females that were sampled on any one day were paired on the basis of proximity in age of their infants. Such pairing provided the possibility of control for variability in factors such as day journey length or unusual daily events and enabled me to sample females more frequently than if I had followed only one female per day. Alternating between two females did have the disadvantage that more time was devoted to searching for individuals and probably more of the 15-minute samples were missed altogether than would have been the case if I had stayed with just one female each day.

At the beginning of each 15-minute sample period and at the end of each of the three 5-minute intervals thereafter, I took point (instantaneous) samples (J. Altmann 1974), recording: (1) the mother's behavior state (categorized as: feeding, walking, engaged in a grooming interaction, engaged in other social interactions, or resting), as in Slatkin (1975, Slatkin and Hausfater 1976); (2) the identities of all the mother's neighbors within 2 meters and those more than 2 but less than 5 meters away; (3) the distance between the mother and her infant (categorized as: in contact, within mother's arm's reach but not in contact, within 2 meters but at greater than arm's reach, 2 to 5 meters, 5 to 10 meters, 10 to 20 meters, or greater than 20 meters); and (4) whether the infant was playing with or in contact with any other individual. The point-sample data were obtained to provide unbiased estimates of the percentage of time spent in various states or activities for which I was unable to keep continuous records. In particular, the point samples provided the ecological information on maternal time budgets, neighbor relationships as part of characterization of the social milieu, and the degree of coordination between a mother's behavior and infant contact with its mother.

The entire 15-minute period during which these point samples were taken served also as a focal sample (J. Altmann 1974) taken on the mother, during which I recorded all occurrences of social behaviors (Appendix 4) and all changes in mother-infant spatial relations, categorized as contact, within mother's arm's reach but not in contact, and beyond arm's reach. Onsets and terminations of grooming and of spatial states were timed, and actor and object identity were recorded. Timing was done with a digital electronic stopwatch to an instrumental accuracy of 0.01 minute. For other interactions, only the behavior and partner identity were recorded, with sequential order but not time of occurrence retained in the record. These data enabled me to examine the sequential and behavioral details of social relationships and the dynamics of the development of infant independence.

Each sample period on a mother also served as a focal sample on her infant to the limits of my capabilities. I was able to record all the infant's interactions as long as mother and infant were within 5 meters of each other. However, as the distance between the two increased beyond that, my ability to obtain complete records decreased, particularly for subtle or fleeting behavior and in the woodland parts of the habitat. I originally planned to divide my time between focal samples on mothers and on infants, but this resulted in inadequate sampling of both. Therefore, I restricted myself to staying with the mother when the two separated. This was not an appreciable restriction for very young infants; but a more complete view of the world of older infants awaits further study.

At the end of each 15-minute observation period, I recorded the "predominant activity" of the group and of the focal animal during that observation period. This was a subjective determination of the activity (feeding, walking, resting, or socializing) that occupied more of the time of more of the individuals than did any other activity. The goal was to obtain easily collected data from which I could deduce the existence of synchrony between the focal animal and the rest of the group.

A full day of scheduled samples yielded four-and-a-half to five hours of focal samples plus approximately another five hours of ad libitum sampling, photography, and map work. Table 3 lists the in-sight observation time for focal samples on mothers starting with 1 August 1975.

In addition to the formal sample periods, necessary search time, and occasional out-of-sight periods during samples, an approximately equivalent amount of time was spent, either on sample days or on additional days, collecting various other data relevant to this study in particular or to the longitudinal project in general. Thus, a total of approximately 1,800 hours was spent in close observation during the study. The group's day route was plotted on a 1:14,761 map and the location of the center of the group was recorded on the hour and the half hour (S. Altmann and J. Altmann 1970). Predation on or by members of the group and the occurrence of wounds other than small scratches were also recorded as part of our longitudinal project and as background

Table 3. Basic data for each mother-infant pair. The observation time given for each month is the actual time-in-sight. Months marked with an asterisk were not used for all analyses owing to incompleteness at the beginning of the study. Infants are ordered by birth dates.

Infant Name	Birth date	Health at birth[a]	Mother Name	Parity[b]	Pregnancy	1	2	3	4	5	6	7	8	9	10	11	12
♀ Pooh	31 Oct. 74	W?	Plum	P	0	0	0	0	0	0	0	0	0	0	154.04	140	120
♂ Ozzie	24 Dec. 74	H?	Oval	M(≥4)	0	0	0	0	0	0	0	0	180	115	0	0	0
♂ Fred	1 Jan. 75	H?	Fem	P	0	0	0	0	0	0	0	0	180	180	180	0	0
♀ Alice	1 Jan. 75	H?	Alto	M(≥4)	0	0	0	0	0	0	0	0	180	120	105	0	0
♀ Eno	22 Apr. 75	H?	Este	M(≥4)	0	0	0	0	250	303.91	255	270	150.29	135	0	0	0
♀ Summer	7 July 75	H?	Scar	M(≥3)	0	269*	293.58*	408.89	240	270	115	135	150	165	0	0	360
♂ Pedro	9 July 75	H?	Preg	M(≥4)	0	170*	180*	329.22	245	255	140.20	150	135	0	0	0	0
♀ Misty	15 Aug. 75	W	Mom	M(≥4)	180	620.40*	540	135	0	0	0	0	0	0	0	0	0
♂ Bristle	17 Aug. 75	H	Brush	P*	160	621*	435	375	465	480	335.88	150	270	390	135	240	135
♂ Hans	15 Oct. 75	H	Handle	P	135	775	660	540	471.30	375	345	390	465	405	120	0	240
♂ Grendel	27 Jan. 76	H	Gin	P	240	907.28	540	345	657.58	525	535	0	0	405	0	0	0
♀ Sesame	2 Feb. 76	W	Slinky	P*	240	675	675	345	525	660	390	0	0	405	0	0	0
♂ Juma	28 Feb. 76	H	Judy	M(≥4)	240	595	435	255	0	0	0	0	0	0	0	0	0
♀ Safi	2 Mar. 76	H	Spot	P	225	780	405	750	450	493.32	0	0	375	0	0	0	0
♂ Moshi	1 June 76	H	Mom	M(≥5)	270	735	525	0	0	505	0	0	0	0	0	0	0
♂ Oreo	28 July 76	H?	Oval	M(≥5)	270	0	0	525	0	0	0	0	0	0	0	0	0
♀ Vicki	4 Oct. 76	H?	Vee	P	0	650	0	0	0	0	0	0	0	0	0	0	0
♀ Peach	8 Oct. 76	H	Plum	M(2)	195	615	0	0	0	0	0	0	0	0	0	0	0
Total focal sample in-sight time					2,155	7,143.68	4,688.58	4,008.11	3,303.88	3,867.23	2,116.08	1,095	2,085.29	2,140	694.04	380	855
Total individuals sampled					10	11	10	10	8	9	7	5	9	8	5	2	4

a. Infant health (H = healthy; W = weak) was judged on the basis of clinging ability, vigor, skin appearance (bright red as opposed to pale pink or gray-toned), coat, (thick and shiny considered healthy), eyes (clear considered healthy). Health assessments took several days of close observation. For those infants for whom this was not possible, assessment was based on the available information from more casual or brief observations and marked with a ?.

b. Maternal parity is marked P for primiparous, P* if all previous pregnancies resulted in stillbirths or neonatal death (see Appendix 1), M for multiparous with actual parity indicated in parentheses.

ecological and group data within which to examine the maternal experience.

Several kinds of data were collected on an ad libitum basis (J. Altmann 1974). First, identities of partners in sexual consortships were recorded for later comparison with adult male relationships with mothers and the infants that were conceived at these times; that is, I made an attempt to ascertain likely paternity. Such a field determination, though not ideal is more reliable in baboons than in many other primates because sexual consortships are relatively obvious, enduring, and exclusive mating partnerships and because timing of ovulation is well correlated with (one to three days preceding) onset of deturgescence of the sexual skin swelling (see Gillman and Gibert 1946, Kriewaldt and Hendrickx 1968).

Second, records were kept of the outcome of agonistic bouts in the group as a whole. An agonistic bout was defined as an interaction sequence between two individuals in which one exhibits at least one "submissive behavior" or "aggressive behavior" (see Appendix 4). Dominance was then assessed by the criteria of Hausfater (1975a): "Winner-loser determinations were made on the basis of the behaviors given during agonistic bouts. A winner and a loser were determined in an agonistic bout only when one animal directed one or more submissive behaviors, and no aggressive behaviors, toward a second animal who directed no submissive behaviors toward the first animal. The individual who gave the submissive behaviors was considered the loser of the bout, and the individual who gave only aggressive and/or other nonsubmissive behaviors was considered the winner of the bout" (Hausfater 1975a:25). Dominance relationships so defined are stable over periods of several years, often a whole lifetime, for adult females, whereas among adult males the relationships often change within a year (Hausfater 1975a and Hausfater et al. in preparation).

Third, I recorded grooming of juveniles by adult females, in order to examine long-term affiliative relationships.

Fourth, I recorded the identity of individuals other than the mother that carried any infant, and the circumstances under which this occurred.

Fifth, I recorded the circumstances under which infants gave distress calls, particularly times when they "threw tantrums" (described in Chapter 9). By reducing the types of ad libitum observations to a select few, I was able to concentrate on obtaining more complete records of those few than is possible with a larger set.

Additionally, some ad libitum descriptions of mothers and of infants were made. Also, at least once a week during the first few months of infant life, I made notes on each infant's physical and behavioral development. Such descriptions were less frequent during later months unless particular developmental milestones were reached or unusual events such as illness occurred.

To obtain demographic and reproductive data, each morning I took a "census" of the group indicating the presence (or absence) of each group member and the presence of any new members, any new wounds or pathologies, and each female's reproductive state: the size and turgescence of the sex skin swelling, and presence (or absence) of pink coloration of the paracallosal skin, indicating pregnancy (S. Altmann 1970).

I stayed with the group from 0730 or 0800 in the morning until 1730 or 1800 in the evening. The animals were never fed, and I tried to minimize my effect on them. As in all work on the group, I tried to be a neutral, noninteractive part of the animals' habitat—a nonparticipant observer. As is usually done in behavioral field work, data were collected by a single observer. A number of actions were taken to improve the reliability of data gathered this way. First, as indicated in Appendix 4 and above, the categories that I recorded are relatively noninterpretive descriptions of gross motor patterns. Second, as noted previously, I did not use data from early observations despite the fact that I had previously conducted a number of other studies of baboons. Each new field study requires a period of adjustment and practice. Additionally, I had several periods of informal reliability checks with Glenn Hausfater and Stuart Altmann for the most difficult behavioral categories, such as glancing and subtle spatial displacements, as well as for estimation of distances. For distance estimation, initially, I also repeatedly practiced by tossing out objects, estimating the distance between them, and then measuring the distance.

Data Processing

During the first months data were recorded on a Sony BM 10 cassette recorder and subsequently transcribed. Once I became practiced, the behavior categories became well established, and the sampling became routine, the advantages of recording directly on paper outweighed the disadvantages. To a considerable extent this was made possible by working on foot rather than from a vehicle as we had done previously, and also by the recent development of small electronic digital-readout stopwatches. Such watches are markedly easier to read and more accurate than traditional clock-face stopwatches.

My field notes were then entered on the IBM 370 computer on the University of Chicago campus. The time of each timed event was recorded, followed by the behavioral record. Untimed events were entered in the same format, each with a sequence number rather than a new time. Each record included the actor, recipient, and behavior(s) by that actor. Little coding or transformation of the data was required thanks to the availability of CRESCAT (Kaye 1977), a program system that was developed by Ken Kaye, Steve Muka, and Starkey Duncan at the University of Chicago. CRESCAT has considerable capability for editing, sequential pattern searches, and handling of matrices, and it is particularly suited to entering and analyzing the kind of data I gathered. I then wrote programs in the CRESCAT language to extract and analyze subsets of the data files. (Details of the data processing are available from the author on request.) In addition, I created a separate file consisting of a single record for each female on each sample day. This record included the infant's age and the length of the group's day journey as well as data obtained from CRESCAT analyses, such as the percentage of time an infant spent in contact with its mother. These records were then entered as case records in SPSS (Statistical Package for the Social Sciences; Nie et al. 1975) for analysis of ecological data and infants' daily contact time.

Glance Rate

During July and August of 1974 I used focal animal sampling to provide estimates of the rates of visual glancing by adult females of Alto's Group. Glancing is probably the major means by which a baboon monitors its social and physical environment (Chance 1967, Chance and Jolly 1970). I suspected that individual differences in glance rate would reflect differences in tension or attentiveness to the external world and might be reflected in maternal care. Accurate recording of glances requires exceptionally good observational conditions and intense concentration. I therefore placed a limit of ten minutes on each of the samples that I collected and eliminated the concurrent collection of other behavioral data. I anticipated that glance rates would vary between the two major habitats occupied by this group, woodland and grassland, so I attempted to collect at least three ten-minute glance samples for each of the 14 fully adult females in each of the habitats. All visual glancing was included, whether aimed at conspecifics or at other aspects of the environment, except that looking at food sources immediately next to the animal was not included. Unfortunately, I found that I could not reliably tell at whom an animal glanced except for most instances of glancing between mothers and their infants. Even the latter data are much less complete than records of objects of other behaviors and could only be scored when done by (rather than to) the focal animal unless the subject and object animals were near each other.

Glance samples either were not taken, were discontinued, or were later discarded under the following conditions: (1) If the subject animal was glancing disproportionately at me. (2) If the predominant activity of the subject animal was walking. Glancing was too difficult to observe in this situation, and I was forced to follow an animal to an extent that was usually disturbing to it or those around it. (3) If the subject animal was engaged in a grooming interaction. During grooming interactions glance rate of both groomer and groomee drops virtually to zero. Inclusion of such samples would have unduly biased the results in a study as short as this (1974, two months). Ideally, large samples systematically distributed through the day and taken for many days would include such activities (grooming and walking). (4) If a predator-alarm was in progress in the group or if the subject animal was in the midst of a fight. These circumstances were excluded owing to the extreme difficulty of recording the high glance rate that occurred in these difficult situations, as well as because of the undue effect the inclusion of a few such samples for a few individuals would have on the resultant data set, as described in (3) above.

Sample Sizes

A study such as this presents a number of problems of sample size that warrant comment here. Sample sizes in this study are for the most part larger than have been available from systematic research on any wild, unprovisioned primate. For some analyses they are larger than published data, even for provisioned or caged primates, or they deal with topics that cannot be or have not been tackled for such groups. However, the sample sizes are neither uniformly large nor uniformly small for the comparisons I make. For example, even though the demographic data span seven years rather than just the 15 months of the mother-infant study, they are for just one social group, and sample sizes for infant survivorship are now appreciable in total but not when partitioned by infant gender, maternal parity, and dominance rank. Likewise, seven years of data provide relatively small samples regarding mortality for adults and older juveniles. For the behavioral data during the mother-infant study, sample sizes are extremely variable from one topic to another, partially because of factors such as infant mortality or less intensive sampling of older infants. Samples are small in other situations because some events, though probably quite important, are also very rare. Another source of small samples is the fact that this is a developmental study, and that mothers and infants pass through some crucial periods very quickly—the extreme case being the first day of infant life. Of course, for some analyses I have very large samples of behavior for many infants, in some cases much larger samples of individuals and behaviors than have resulted from comparable laboratory investigations.

Rather than either discarding all small samples or, conversely, retaining all samples and treating them as if they provided equally strong evidence, I have tried throughout the remaining text to indicate the data base on which various statements rest. I tried to do this without belaboring the issue; my hope is that alerting readers at this stage will enable them to read the following chapters with the necessary caution yet without being too encumbered by repeated detailed warnings. I have usually tested ideas that seemed most straightforward or compelling from a biological standpoint. I have then applied one-tailed statistical tests, but usually provided the raw data as well.

In the analyses and presentation of the data that follow I have taken what might be considered a conservative approach. Unless otherwise noted, for each analysis I have treated all data for a mother-infant pair during each month of infant age as a single data point. Thus, it is the average over individuals and the standard error of that mean that are graphed. Had I treated each act or even each day's data as the sampling unit, sample sizes would appear much larger and standard errors usually much smaller, but there would be questions of independence of the points that contribute to the calculations for each month and problems of undue contribution of particular individuals, effects of daily ecological variance, and so on. Above each graph, I indicate the sample size, both the number of individuals who contribute to each month's data and the number of acts or amount of observation time that contribute to that data.

Additional details of methods and analysis are provided as appropriate in relevant sections.

4 / Demography: Births, Deaths, and Interbirth Intervals

AGE DIFFERENCES BETWEEN SIBLINGS, the size and age-sex composition of available play groups, and other important features of the social milieu of mothers and infants are determined by the demographic patterns of a population—age-specific rates of natality and mortality, migration, and so on. Mortality figures in particular provide clues to stressful life stages and are important in considering costs and benefits of various social behaviors within an evolutionary framework.

Survival of Infants, Juveniles, and Adult Females

Of the 13 infants who were born during the study, 3 (Pedro, Misty, Vicki) died during the study after varying periods of illness, 1 (Juma) suddenly disappeared overnight with his mother during the study, 1 (Safi) became seriously ill and recovered, and 3 (Sesame, Moshi, and Peach) died after the study but during the first two years of life, that is, as infants (first 12 months) or in the beginning of the juvenile period. Six (Summer, Bristle, Hans, Grendel, Safi, and Oreo) survived the first two years of life, 2 of them (Bristle, Oreo) as orphans. Of the 5 infants who were included in the study but who were born before July 1975 (i.e., had already survived part of the first year when the mother-infant study began), 4 (Alice, Ozzie, Fred, Eno) survived through the second year of life and 1 (Pooh) died just before her second birthday. Alice was orphaned before her second birthday; Ozzie after his.

Only rarely can we determine either the immediate cause of death or an earlier precipitating cause. When an apparently healthy female or young juvenile disappears from the sleeping grove overnight we presume not only that it died but that it was probably taken by a predator (but see Juma, Appendix 2 and below). However, in the case of older juvenile, subadult, and adult males even overnight disappearances can be due to migration rather than death, a fact that makes it difficult even to construct life tables for these age-sex classes. Determination of an infant's death is somewhat easier because if it is survived by its mother, she usually carries the corpse for several days. Of the infants in the 1975–76 study Peach and Misty apparently died of disease; probably Pedro and Juma did as well, along with Juma's mother, Judy (Appendix 2). Vicki's death was probably due to poor mothering, and Pooh's to the severe wounds inflicted by male Even just before her second birthday (Appendix 2). Both Moshi and Sesame showed signs of possible nutritional deficiencies (see Chapter 8 and Appendix 2) for at least a year before they died. Sesame disappeared just before her second birthday and soon after the birth of her mother's next infant. Moshi fell from a tree when his younger brother was a few months old. The cause of his fall is unknown. Alice's mother, Alto, disappeared while we were away from the group for several days. When we first identified Alto in 1969, she appeared to be old, probably at least 15 years of age; during the last months of her life she seemed to become less interactive and to rest more.

Mortality patterns during the year of the mother-infant study were similar to those for the whole seven-year period of the longitudinal study; we can analyze the latter in greater detail because of the larger sample sizes. Fig. 9 provides survivorship data for the 54 individuals born into Alto's Group from July 1971 through July 1978. The complete life-table data appear in Table 4. Sample sizes are now about twice those available in our earlier paper on infant survivorship (J. Altmann et al. 1977), and data are now available for years two through five, taking females through the age of menarche and males almost to the beginning of subadulthood. In addition, the mortality data considered here come entirely from a period of fairly stable demographic parameters. No trends in survival could be detected when I examined infant mortality within Alto's Group from each year separately, and I have therefore combined the data since 1971 to provide more adequate sample sizes for the analyses that follow.

Several features of the survival data are quite striking. Mortality is appreciable (.28) in the first year of life and only slightly less so (.25) in the second year. However, it is essentially zero in the next three years. (One juvenile male, Kub, either died or migrated in the fourth year.) Baboon mothers provide little care during an offspring's second year of life, often none during the latter part of that year. The low survival rates during the second year of life suggest that a prolongation of maternal care might potentially be of considerable benefit to these offspring and increase their rates of survival. By contrast, in the third, fourth, and fifth

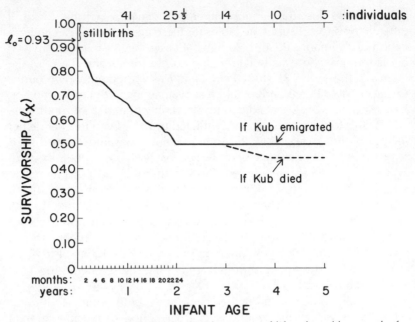

Fig. 9. *Survivorship curve for the first five years of life, plotted by months for the first two years and by years for the following three years. For each age, the point plotted is l_x, the probability of surviving to that age. The initial sample is all full-term pregnancies, that is, l_0 is 1 − stillbirth rate, or .93. To obtain the more conventional curve for just live births, multiply each value in the present graph by $1/.93 = 1.08$. The separate lines after the third year result from the disappearance of one juvenile male (Kub), which may have been due to either death or migration.*

years, which are still prereproductive years, the survival rate is virtually 100 percent. Consequently, it seems unlikely that directing any additional care or "altruistic" behaviors of immediate consequence toward these older juveniles would result in a selective advantage: there is no potential survival difference that could provide the selective basis for the evolution of such behaviors. However, behaviors of long-term or delayed consequence, such as those affecting age of menarche, ability to obtain mating partners, or ability to learn parental skills, may be of advantage to this age group.

In early research (e.g., DeVore 1963, S. Altmann and J. Altmann 1970), the period of infancy was considered to be the first year of life. However, from our longitudinal study we have since found that individuals identified as older infants by physical appearance and by the fact that they are still suckling and are associated with a single adult

Table 4. Survivorship of infants born into Alto's Group,
July 1971–July 1978.

Age interval $[x, x+1]$	Sample size N_x	Mortality rate q_x	Survival rate $p_x = 1 - q_x$	Survival from full-term pregnancy through age $x + 1$ $l_x + 1 = l_0 p_0 p_1 \cdots p_x$ $l_0 = 1 -$ rate of stillbirth $= 0.93\ (N = 54)$
In months				
[0,1]	49	.08	.92	.86
[1,2]	45	.02	.98	.84
[2,3]	44	.05	.95	.80
[3,4]	41	.05	.95	.76
[4,5]	39	0	1.00	.76
[5,6]	38	0	1.00	.76
[6,7]	37	.03	.97	.74
[7,8]	36	.03	.97	.72
[8,9]	33	.03	.97	.70
[9,10]	31	.01	.99	.69
[10,11]	30.83[a]	.01	.99	.68
[11,12]	30.66	.01	.99	.67
[12,13]	28.33	.05	.95	.64
[13,14]	27.00	.01	.99	.63
[14,15]	25.67	.01	.99	.62
[15,16]	25.33	.05	.95	.59
[16,17]	23.17	.01	.99	.58
[17,18]	23	0	1.00	.58
[18,19]	22	0	1.00	.58
[19,20]	22	0	1.00	.58
[20,21]	22	.05	.95	.55
[21,22]	21	0	1.00	.55
[22,23]	21	.05	.95	.52
[23,24]	20	.05	.95	.49
In years				
[0,1]	41	.28	.72	.67
[1,2]	25.33	.25	.75	.50
[2,3]	14	0	1.00	.50
[3,4]	10	.10 or 0	.90 or 1.00	.45 or .50
[4,5]	4	0	1.00	.45 or .50

a. In two cases death was not known to the nearest month. Those were prorated over the months of life during which death might have occurred.

female, are up to 16 months of age (J. Altmann et al. 1977). Baboons are not strictly seasonal breeders, interbirth intervals are individually variable, and the development of independence is a long, gradual process that extends into the second year of life. Thus it is difficult and somewhat arbitrary to define the age limits of infancy. All of the first year of life is clearly a period of dependence and infancy. This study focuses on that period and it is individuals of that age that I refer to as infants. However, considerations such as the mortality data above and mothers' future reproduction (J. Altmann et al. 1978, and see below) demonstrate the importance of the second year of life, and where possible, I extend my discussion to this period.

Consider next mortality risks for females during adulthood. In the seven years since mid-1971, we have data for approximately 78 female-years of exposure for fully mature females, defined as those who have conceived at least once. Nine fully adult females died in Alto's Group during the period from September 1971 to September 1978, for a mortality rate of .12 per annum. In most mammals mortality risk is low in the early years of adulthood, but rises, often rather sharply, in later years (Caughley 1966). Is mortality risk a function of age among adult female baboons? In particular, is the mortality risk to mothers age-related? Young adult females (taken here as those who have conceived at least once but who were within three years of first conception, that is, those who were approximately six to nine years old) had an annual death rate of .04; older ones (defined as those who were at least three years past first conception) a rate of .16. Adult females usually spend approximately one-half of their adult lives with a dependent infant under a year old—the proportion is slightly lower if the infant dies, higher if the infant survives. Thus, one would expect mothers of infants to account for half of the deaths among adult females. In fact, six of the nine females who died had infants under a year of age at the time of death, for a rate of .15 deaths per annum for females with an infant. The rate for females without an infant was .08 per annum. Of these remaining three deaths, one was a young female (Brush) experiencing her third pregnancy. The other two were much older females (Jane, Alto) who were experiencing abnormal reproductive cycles: in Jane's case, a particularly long cycling time without becoming pregnant, and in Alto's an unusually long postpartum amenorrhea (Fig. 4). Perhaps their deaths resulted from or were made more likely by pathology or old age. The tentative conclusion is that in this population reproduction exacts a substantial cost in the form of increased mortality.

It is reasonable to expect that two additional variables might affect the survival of mothers or infants: maternal dominance rank and infant gender. In the past seven years, sex ratios at birth have been equal (26 males : 26 females, plus two unsexed stillbirths and one unsexed case of neonatal death during late 1972). Infant (first-year) mortality, including stillbirths, was higher for male infants than for female infants (10 out of 23 for males versus 6 out of 21 for females). This difference does not approach statistical significance ($P > .10$, Fisher Exact Test), and was due almost entirely to the fact that four of the male infants and only one of the female infants died at the same time that their mothers did. We cannot yet tell whether this latter result is an artifact or whether, for some reason, mothers of male infants are more vulnerable. Although these mothers were all older mothers, older mothers, like younger ones, produced infants at about a 1 : 1 sex ratio. For humans, sex ratios at birth are slightly skewed toward males and become less so for older mothers or those of higher parity, but a sharp skewing in favor of males occurs for quite old mothers (Hytten and Leitch 1964, Teitelbaum et al. 1971). We do not yet have sample sizes for the baboons adequate to detect differences of the small magnitude found in humans.

It is generally assumed that female dominance rank is correlated with reproductive success. Although this has not yet been demonstrated for savannah baboons, there is some positive evidence for provisioned macaques (Drickamer 1974, Sade et al. 1977) and perhaps for geladas (Dunbar and Dunbar 1977). Female offspring assume their mother's relative dominance rank for life, whereas male offspring repeatedly change rank during adulthood (Hausfater 1975a) and seem to be less affected, as adults, by their mother's rank (Hausfater et al. in preparation), as has been reported for macaques (e.g., Kawai 1958, Sade 1967).

As predicted on the basis of the behavioral life history results described above, in Alto's Group dominance rank is related to sex of offspring: low-ranking mothers produce more male than female offspring, high-ranking ones producing primarily female offspring (Table 5). We can make a dichotomous characterization of these mothers as high- or low-ranking. Then we can also make a dichotomous characterization on the basis of whether a female had more male or more female offspring, considering each mother as the sampling unit, rather than each birth, in case there is a lack of independence of sex among offspring of the same female. The results (omitting ties) are that five or seven high-ranking females had more female than male infants, and five of seven low-ranking females had more male than female infants. Alternatively, if we do consider each birth as an independent

Table 5. Sex of infants born in Alto's Group with females ordered by dominance rank at conception, July 1971–July 1978.

Female	No. ii of unknown sex	No. ♂ ii	No. ♀ ii	Total no. of ii	Majority sex
T.T.	0	2	1	3	♂
Alto	0	0	2	2	♀
Spot [a]	0	0	2	2	♀
Mom	0	3	3	6	=
Fluff	0	0	1	1	♀
Vee	0	0	2	2	♀
Preg	0	2	2	4	=
Scar	0	1	3	4	♀
Oval	0	3 [b]	2	5	♂
Ring	0	1	0	1	♂
Fem	0	2	1	3	♂
Gin	0	1	1	2	=
Jane	1	1	0	2	♂
Slinky	1	1	2	4	♀
Handle	0	2	0	2	♂
Plum	0	1	2	3	♀
Brush	0	2	0	2	♂
Este	0	3	1	4	♂
Judy	1 [c]	1	1 [c]	3	=
Total	3	26	26	55	

a. When Spot was first identified as a yearling in 1971, she had a small but visible penis and femalelike callosities (divided at the midline). We classified her as a male (Hausfater 1975a) until mid-1974, at which time we could no longer see a penis and she began sexual cycles (J. Altmann et al. 1977).

b. When Oval conceived the third male infant, she had dropped in rank to a place between Slinky and Handle.

c. When these infants were conceived, Judy ranked just above Ring. Note that the changes for infants of Oval and Judy would improve the relationship between maternal rank and infant gender, i.e., the tendency for higher-ranking mothers to produce female infants, lower-ranking ones to produce male infants.

sample and incorporate the two rank changes, we see that 10 of 29 infants born to high-ranking females were male, but 15 of 22 infants born to low-ranking females were male ($P < .02$, One-tailed Fisher Exact Test). Although I know of no literature on the possible mechanism leading to this result, physiological differences due to differences in stress levels could perhaps result in differential sperm survival in the two groups. The observed sex difference of offspring would appear to be the best strategy for each type of female, if female dominance rank is correlated with reproductive success, with female offspring of high-ranking mothers retaining their mother's rank and male offspring of low-ranking females being "freed" of their mother's rank. The relationship between maternal rank and sex of offspring has been modeled by Trivers and Willard (1973), and the model I am proposing for the observed baboon sex ratios is in the same spirit. However, this social and reproductive system requires a different set of assumptions than those used by Trivers and Willard; and therefore different specific predictions result (see also Clark 1978 and Myer 1978).

From the results that survival rates were as good or better for female (versus male) infants, and that high-ranking females tended to produce female offspring, one would expect to find that offspring of high-ranking females have lower mortality rates than those of lower-ranking females. However, rates of infant mortality were slightly but not significantly lower for infants of low-ranking mothers (6 out of 20 = .30) than for infants of high-ranking mothers (9 out of 23 = .39). Thus despite the facts that low-ranking mothers produce more male than female offspring, high-ranking ones produce primarily female offspring (Table 5), and female infants have survival rates as high as or higher than those of male infants, infants of high-ranking mothers have no higher, and perhaps somewhat lower, survival rates. This contraintuitive result for infant survival rates contradicts the limited available primate data on reproductive success from two provisioned rhesus colonies (Drickamer 1974, Sade et al. 1977), as well as data on survival of human infants in different socioeconomic classes (Hauser and Duncan 1959). However, even though infant survival does not appear to be positively correlated with dominance rank, other factors affecting the female's reproductive success, such as age of menarche and length of interbirth interval, may be rank related in our population. Only future study will clarify this puzzling situation.

When we combine the various pieces of information on maternal and infant mortality, the suggestion emerges that young adult females, essentially those with their first or second infant, are about as likely to have that infant survive and are more likely to survive themselves than are older females. Older females are more likely than are younger ones to die during the period of infant care, in which case the infants will also die. However, if these older mothers survive, their infants are also very likely to survive. Thus, their fate, more than that of young mothers, is closely linked to that of their infants during the first year.

Perhaps a cautionary reminder is necessary at this point. Some of the foregoing results have been obtained by partitioning a small data set, resulting of course in smaller subsets. Also there are several factors that may lead to differential exposure to the various reproductive stages, including the very results just discussed. There is at present no way to remove such confounding completely. Patterns will, one hopes, become clearer as more data accumulate. Sample sizes are not yet adequate to confirm statistically even appreciable differences.

Interbirth Intervals

Not only do infants affect their mothers' survival chances, but infants additionally affect demographic processes in a primate group in general and their mothers' reproductive success in particular through the direct effect that infants have on their mothers' future reproduction (see J. Altmann et al. 1978 and references therein). Mothers of surviving Amboseli infants experience approximately 12 months of postpartum amenorrhea and then take an average of four cycles to conceive, whereas infant death results in resumption of cycles within one month of the death and conception after only one or two cycles on the average (P < .02; see preceding reference for details). It has been further suggested that mothers could improve their reproductive success if they reduced infant care, for example, by weaning their infants earlier (e.g., Trivers 1974), thereby reducing the length of postpartum amenorrhea. While this has certainly been the case among humans in developing countries in recent years (see Hauser and Duncan 1959, Lee 1978), such a result is dependent on (1) infant mortality rates not being appreciably increased by early weaning and (2) the mortality risk of childbirth and early stages of infant care being low. Weaning foods and even a semblance of modern medicine are probably sufficient to satisfy these conditions in developing countries. It is unlikely, however, that these conditions prevailed until recently for humans, or that they are satisfied in most animal habitats, including that of the Amboseli baboons. It is quite possible that in the absence of such advantages of early weaning females obtain higher reproductive success by engaging in fairly long periods of infant care than they would by reducing the period spent caring for a current infant in order to reduce the interbirth interval.

Summary

A per annum mortality rate of almost .30 occurs during each of the first two years of life, with mortality dropping to virtually zero in the several years thereafter. Thus infancy and the following year are particularly crucial and difficult periods. The data also suggest that older adult females suffer higher mortality rates than do younger ones. This general pattern is the common mammalian one (Caughley 1966): only the sharp discontinuity at age two may be unusual. Dittus' study of *Macaca sinica* (1975, 1977) provides the only comparable mortality data for a nonexpending natural primate population. His data exhibit the same U-shaped distribution. Dittus' life-table data come from one cross-sectional census of many troops. Age was estimated by physical appearance, for immatures by making use of the short, discrete birth season in this species, and by comparisons with longitudinal changes in a smaller set of known individuals studied for several years. The life-table data that Dittus presents (1975:132–138) agree in general with those reported here, with high rates of mortality for infants and for older adults, low rates for older juveniles and young adults. The one striking difference that merits comment here is the much higher mortality rate that Dittus reports for young female versus male infants at his site in Polonnaruwa. However, this finding is based on the assumption that the 1970–71 sex ratio at birth was 1:1. Although known births for four years averaged a 1:1 ratio, the ratio reported for the Polonnaruwa region for 1970–71 is 21:14, or 3:2, essentially the same ratio found in the infant-2 (older infant) class during the census. Thus it is not clear from the available data that differential female mortality occurs among the infants of that population. Subsequent data from the longitudinal project there should clarify this issue.

Additionally, in Amboseli there is a cost of reproduction. That is, females, especially older females, with dependent infants tend to suffer higher mortality rates than do younger ones or those in other stages of the reproductive cycle. I know of no published data on the cost of reproduction in other wild primate populations. Infants may impose an additional cost on their mothers' future reproduction due to an extended period of postpartum sterility, but this will depend on the risks mothers and infants would incur as a result of early weaning. Thus mortality risks are high for both mothers and infants.

In the following chapters I shall attempt to identify those ecological and social factors that shape the first two years of life and in particular those that may enhance or hinder the survival of mothers, infants, and young juveniles and affect their interactions.*

* Deaths of adult females (four) and births (six to multipara, five to primipara) from 1 August 1978 through 31 December 1979 provide increased evidence for (1) the susceptibility of adult females to death when they have infants (p. 36) and (2) the relationship between maternal dominance rank and infant gender (pp. 37, 38).

5 / Ecology and Maternal Time Budgets

IN RECENT YEARS time and energy have figured prominently in minimization or maximization models in behavioral ecology (e.g., S. Altmann and Wagner 1978, Pyke et al. 1977, Krebs 1978). However, the importance of time may be much more general than is suggested by minimization models. All of an animal's basic activities require time to perform; some of them require considerable amounts of time. Moreover, time is quite limited and all activities must be carried out within a fixed number of hours: baboons spend all 10 of the dark nighttime hours resting and sleeping in the trees, plus 3 to 4 additional late-evening and early-morning hours. Ten to 11 hours remain. I suggested previously that these daytime hours are filled to a considerable extent by maintenance activities. As a key to understanding the demographic results of the preceding chapter and the constraints within which mothers live, I shall now consider these daytime time budgets in detail and the ways in which the presence of a dependent infant influences a mother's productive activities.

Seasonality of Time Budgets

The Amboseli habitat is subject to considerable seasonal variability. As expected from its latitude (2°40'S), there is little variability in day length and temperature range (Hausfater 1975a, Post 1978, Western 1973), but seasonal variability in rainfall is considerable (Fig. 10). The period from May or June through October is usually characterized as the long dry season, November and December as the short rains, January and perhaps February as the interrains, and March through April or May as the long rains (Western 1973). Despite some variability from year to year, this basic pattern persists in Amboseli (Hausfater 1975a, Post 1978, Western 1973, unpublished data). During the dry

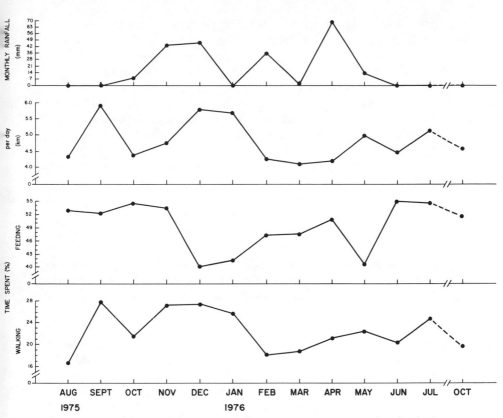

Fig. 10. *Monthly variability in rainfall, group movements, and individual's time budgets. Rainfall records were kept at a base camp by David Klein and Wesley Henry during 1975–76.*

season the baboons depend on the few permanent waterholes in the central area of the park (S. Altmann and J. Altmann 1970). When the rainy season arrives, however, temporary rain pools are available throughout their home range. At that time most large mammals in the area migrate out of the park (Struhsaker 1967, Western 1973), while others, such as the baboons, alter their habitat usage and movements within the same basic area (Post 1978). During 1975–76 the distance traveled by the baboons was greatest (almost 6 kilometers daily) in December, January, and September. The values were lowest during the rainy season (4 to 4.5 kilometers). Time spent walking showed a somewhat similar pattern, with highs of over 25 percent of the daytime spent walking in September and November through January and lows of 17 to 19 percent in August, February, and March. Variability in feeding time was even greater, with 52 to 55 percent of the time spent feeding during June through November (the rains arrived in late November in 1975) and averages of only 40 to 42 percent during December, January, and May, with a gradual rise after January but a very sharp rise from May to June and for the rest of the dry season.

The effect of rainfall patterns on the group movements and on the percentage of time the baboons spend feeding and walking is a complicated one. Baboons are omnivores (Altmann and Altmann 1970, DeVore and Washburn 1963, Hall 1963, Hamilton et al. 1978, Post 1978, Rowell 1966b), and their various foods are distributed differently throughout their habitat and have different seasonal growth patterns. Particular food plants in particular parts of the animals' habitat leaf or fruit at various times after the onset of the rains. Then, too, some of these plants fruit annually, a month or two after the onset of one of the rains, whereas other plants, such as the grasses, develop lush new growth soon after the onset of each rainy season. Thus, in a linear regression, monthly rainfall accounted for less than 20 percent of the daily variance in distance traveled, and distance traveled or the time spent walking were no better as predictors of the time spent feeding. However, the average distance traveled by the group in a month accounted for 75 percent of the variance in the average time mothers spent walking that month. Also the sum of the rainfall in a month plus that for the previous month accounted for 50 percent of the variance in average monthly time spent feeding. That is, cumulative rainfall was a good predictor of the time an individual spent feeding, and the length of the group's day journey was a good predictor of the time an individual spent walking.

The patterns of seasonal variability in day journey length and in the percentage of time spent feeding and walking that are depicted in Fig. 10 are essentially the same as those found for 1963–64 for day journey length (S. Altmann and J. Altmann 1970) and for 1974–75 for all three variables (Post 1978). Thus it seems reasonable to assume that these environmental conditions represent a pattern that is repeatedly encountered by the baboons from one year to the next, with relatively small annual variations, and that these conditions may be of considerable importance to the energetic requirements of lactating females, to the development of infant independence, and to the survival of both mothers and infants.

Synchrony of Activities within the Group

To what extent is the length of the group's day journey or the amount of time an animal spends feeding determined by any individ-

Fig. 11. *Slinky walking three-legged while clutching Sesame on the day of birth. Note the caked blood on Slinky's perineum.*

Fig. 12. *Judy on the day her infant Juma was born. Juma was born in mid-morning and by mid-afternoon Judy appeared quite fatigued.*

ual and to what extent are an individual's activities determined by those of other group members or the group as a whole? We can assume that neither any individual nor the group as a whole affects short-term rainfall patterns. Yet there remains some amount of flexibility in the amount of time spent feeding, and an individual may influence day journey lengths as well as vice versa. Survival differences may well depend on the extent to which environmental and social constraints combine with reproductive ones to place more or less burden on a female. Likewise, the degree of an individual's ability to affect these variables herself may well make a crucial difference to her survival.

Probably of particular importance to new mothers is the length of the day journey on the infant's first days of life. The infant often has trouble clinging during these first few days, sometimes requiring its mother to walk three-legged while pressing the infant against her ventrum with one hand (Fig. 11), or to perform frequent repositioning during long marches (Rose 1977). Especially on the infant's day of birth, the mother seems quite tired (Fig. 12). This is reflected in her time budget for that day (Fig. 13), during which she spends much more time resting than usual, and in the lack of concordance between her activities and those of other group members, discussed below.

The lengths of the day journey on and near birth days are shown

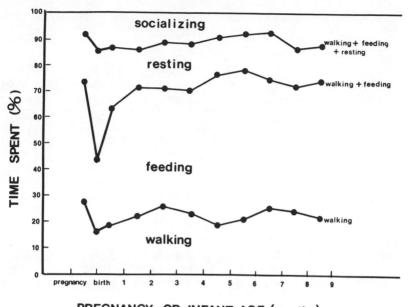

PREGNANCY OR INFANT AGE (months)

Fig. 13. *The time mothers spent in various activities as a function of pregnancy and infant age. Data are plotted separately for the day of birth, but the data for the first month of infant life also include the data for that day. Data are not plotted beyond month nine because data are available for only a few days, and all are from the dry-season months. Note that because an infant's fourth month of life occurs between age three months and age four months, the data for the fourth month of life is plotted midway between those two ages. The same plotting relationship holds for all ages here and in the subsequent graphs in the book.*

in Table 6. Two infants, Juma and Peach, were born during the day journey, rather than at the sleeping grove. For these two the actual day of birth and the next day are both given. As shown in Table 6, day journeys on days of birth were not likely to be shorter than average for the month, despite the mother's fatigue (6/10 lower than the mean, 5/9 lower than the median). Somewhat more of an effect is seen by comparing mean day journey length for the week before the day of birth with that of the week beginning with a birth day. The magnitude of the differences was small, but the direction of the effect was such that in six of eight cases for which I had these data the distance traveled was shorter in the week following infant birth than in the previous week. It remains for further study to determine whether this slight effect is in some sense real (would reach statistical significance with somewhat

Table 6. Length of the day route (km) on and near birth days.

Mother-infant	Birth date	Day of birth	Mean for month including birth	Mean for week preceding birth	Mean for week beginning with infant birth
Mom-Misty	15 Aug. 1975	4.12 <[a]	4.32 (n = 16)	4.26 >[a] (n = 4)	4.20 (n = 5)
Brush-Bristle	17 Aug. 1975	—	—	4.26 > (n = 4)	3.99 (n = 5)
Handle-Hans	15 Oct. 1976	5.00 >	4.38 (n = 13)	5.16 > (n = 4)	3.85 (n = 5)
Gin-Grendel	27 Jan. 1976	4.12 <	5.67 (n = 10)	9.76 > (n = 1)	3.66 (n = 5)
Slinky-Sesame	2 Feb. 1976	2.82 <	4.27 (n = 14)	3.87 > (n = 4)	3.19 (n = 3)
Judy-Juma[b]	28 Feb. 1976	4.67 >	4.27 (n = 14)	4.30 > (n = 1)	3.83 (n = 5)
	29 Feb. 1976	2.50 <	4.27 (n = 14)	4.49 > (n = 2)	3.62 (n = 4)
Spot-Safi	2 Mar. 1976	2.45 <	4.11 (n = 13)	3.82 < (n = 3)	4.13 (n = 4)
Vee-Vicki	4 Oct. 1976	4.72 >	4.57 (n = 17)	—	—
Plum-Peach[b]	8 Oct. 1976	2.45 <	4.57 (n = 17)	4.26 < (n = 3)	4.71 (n = 6)
	9 Oct. 1976	7.54 >	4.57 (n = 17)	3.80 < (n = 4)	4.87 (n = 6)

a. Comparison of values appearing in the two adjacent columns.

b. Infant born during the day journey rather than at the sleeping grove; day of birth and the next day are given for these infants.

larger samples) and if so, whether the magnitude is sufficient to be of any benefit to mothers. New mothers are sometimes seen dragging along at the rear of the group and thereby may be exerting a slight effect on the route, but there is no evidence that the group makes appreciable alterations in its activities to accommodate new mothers. If anything, during the early months of infant independence mothers probably spend more time walking than do other group members, because they spend time retrieving their infants.

I expected that the percentage of time spent feeding, or feeding plus walking, would be a function of maternal dominance rank, with low-ranking females spending more time in these activities than do high-ranking ones. This was predicted on the assumption that low-ranking females would more frequently be displaced from high-quality foods and would therefore need more time to obtain an adequate amount of food. There was no such significant effect on feeding time, nor was there in Post's limited data on this topic (Post 1978). If a small but consistent effect does exist, the data are not yet adequate to detect it. Dominance effects may occur only during the dry season (see "Spatial Displacements" in Chapter 6 and Post 1978). The data from this study could not be analyzed by season owing to the confounding effects of infant age. This remains a problem for future research. Other social constraints on time budgets will be pursued in more detail in the next chapter.

In approximately 90 percent of the samples the focal animal's predominant activity was the same as that of the group. Almost all deviation from that 90 percent figure occurred during the first month of infant life, when there was only 80 percent concordance. Most of this deviation, in turn, occurred on day one of infant life, when there was less than 60 percent agreement. This lack of concordance was due to the fact that mothers fed much less on day one and rested much more. I do not know whether the lower amount of time spent feeding on that day is the result of fatigue and of the difficulty in feeding due to the need to use one arm much of the time for infant support, or whether the afterbirth, which is eaten, supplies sufficient nutrients to allow less feeding that day. On the fifth day of infant life, residual discordance was still produced, partially by new mothers' doing more resting but also by their socializing more, as was true for the rest of month one. The rare instances of lack of concordance in later months were not consistently due to any particular activities.

Effects of Infants on Mothers' Feeding Patterns

In Amboseli the baboons feed while in either a seated or a quadrupedal standing position. Whereas catching insects and feeding on berries are probably more easily done from a standing position, two-handed feeding is done primarily while sitting and may be advantageous for obtaining some foods, such as grass corms (D. Post, personal communication). Although it also appears that feeding while seated is more relaxing and energetically economical, the animals frequently have to move from one food patch to another. Some of these movements are caused by other group members, others by depletion of the present food patch. In either case, the seated animal must stand again with each move. However, during the first few months of an infant's life, if its mother is seated the infant can maintain contact and even doze without clinging and having to support its weight. After many hours of riding on the mother's ventrum, very young infants often tire and have trouble clinging. During the second month the infant can explore with considerable safety within its seated mother's ventral flexure. By the third month an active infant in her ventrum clearly hinders a mother's feeding attempts (see also Rose 1977), and I often observed a mother embrace her infant to passive clinging, push it gently outside her ventrum (and just out of contact but nearby), or stand and feed with the infant exploring under her torso. Mothers of infants that were over five months old stood while feeding slightly more often than they sat (1.11 = mean ratio of the time spent standing while feeding to the time spent sitting while feeding); mothers of three- to five-month-olds were about as likely to sit as to stand (1.01), with considerable variety of direction from month to month and individual to individual. However, during the first month of infant life all nine mothers (P < .01, binomial test) sat while feeding more than they stood (.55) and eight of nine mothers (P < .02) did so during month two (.89) (see also Rose 1977 for similar results in anubis baboons during a short study that did not include the dry season). There were no differences on the basis of maternal dominance rank. Thus a mother's feeding positions seem to be an accommodation to her infant's needs during the first months, but gradually she prevents her infant from hindering her feeding by modifying her infant's behavior and sometimes modifying her own. By month four infants are disproportionately out of contact when their mothers feed.

Maternal Time Budgets and Infant Gender

I can discern no significant effect of infant gender on the time mothers spend feeding or feeding plus walking. However, during all but one month mothers of female infants averaged slightly but insignificantly more feeding time than mothers of male infants. A gender effect resulting in more time spent feeding by mothers of male infants would be expected primarily if male infants grow faster than females. (Adult male baboons are about twice the weight of females.) A differential in the opposite direction would be predicted if males became nutritionally independent on their mothers earlier. The literature is contradictory on the former point for the first year of life for baboon and macaque infants kept in laboratories. Snow's data (1967) indicate no sex difference in growth rates until the end of the second year. Yet for rhesus macaques, *Macaca mulatta* (Van Wagenen and Catchpole 1956), and another baboon colony (Buss and Reed 1970) there are indications that differences in growth rates may exist earlier, with male infants gaining weight more rapidly. During the first year of life, the infants in this study exhibited no consistent visible size differences: female Alice and male Ozzie were approximately the same size; same-aged male Fred was smaller; two-month-older female Pooh was smaller; female Eno, who was four months younger than Alice, Ozzie, and Fred, was about Fred's size; and so on. With respect to earlier nutritional independence of one sex, there were no obvious qualitative differences, but the results of S. Altmann's nutritional study of baboon infants (in preparation) may clarify this issue. I shall turn now to a consideration of the factors that enable an infant to develop nutritional independence.

Maternal Time Budgets and Infant Maturation

We have seen that rainfall and group movements affect individuals' time budgets. How does the presence of an infant create additional demands that may restrict a mother's behavioral options and perhaps precipitate the higher mortality found at this period? In this section I shall consider the effects on a mother's time budget of just one variable, but an important one—the energetic needs of her infant at various ages. The energetic requirements of pregnancy in humans and other mammals are well documented (Blackburn and Calloway 1976a, Brody 1945, Emerson, Saxena, and Poindexter 1972, Hytten and Leitch 1964, Kaczmarski 1966, Naismith and Ritchie 1975). With the birth of her infant a mother begins to provide nutrition through milk production, an energetically less efficient system than the placental one (Reynolds 1967). As the infant grows, its nutritional requirements increase as energy is required to build new tissue, maintain existing tissue, and support higher activity levels as the infant begins to play, explore, and provide its own transportation.

In the discussion of seasonal effects on time budgets I assumed that the differences *between* seasons in the percentage of time spent feeding result not from weight gain or loss or from changes in energetic requirements but rather from differences in feeding efficiency due to changes in food availability. In the present discussion I shall assume that *within* a season differences in feeding time result from differences in energetic demands. Although differences in feeding efficiency probably exist as a result of differences in dominance rank, it seems reasonable to assume that these are slight (Post 1978) compared with differences caused by pregnancy and lactation.

To focus on the major variables and to provide a framework for future refinement, in what follows I explicate a simple algebraic model of the feeding time that a mother would require if she maintained her own body weight and provided all the energy requirements of her growing infant. That is, the basic starting point is an examination of the limits of one aspect of maternal care, provision of the infant's total caloric needs, without immediate detriment to the mother through weight loss. In order to understand the options available to mothers and infants and the factors that might be constraining their behavior as individuals and their relationships as dependent pairs, it is helpful to explore first some limits of the possibilities.

Mathematical simplifications are inevitably somewhat unrealistic (see Cohen 1972), as are nonmathematical simplifications. A mathematical model has the advantage of making simplifications explicit and providing a modifiable framework. Some likely ways of making this model more realistic are given in italics enclosed in brackets, but further refinement of the model itself is not warranted for the present purposes or by the available data.

A Model of Maternal Feeding Time

I shall make the following specific simplifying assumptions:

1. Within any given locale, age-sex class, and season, the percentage of daytime that an animal spends feeding, f, equals $a_0 k$, where a_0 is a constant and k is the energetic requirement of the animal in kilocalories. That is, I assume that feeding efficiency is constant. [*Violations of this assumption probably occur due to dominance rank of the mother (see above), the attraction and interaction of other group members with mothers of young infants (Chapter 6), and the physical presence and movements of an infant near and on its mother (Chapters 8 and 9). A refined model could incorporate this variability.*]

2. $k = a_1 w^{.75}$ where w is the weight of the animal and a_1 is a constant determined by activity level but not a function of weight (Kleiber 1961). That is, I am using an ontogenetic scaling factor of .75 for energetic expenditure. [*Finer analysis might suggest a variable factor ranging from about .66 to 1.00 (see Gould 1975, Schmidt-Nielsen 1977) with the relative expenditure (exponent of 1.00) greatest at the youngest infant age.*] Then, letting $A = a_0 a_1$, we get

$$f = a_0 a_1 w^{.75} = A w^{.75} \tag{1}$$

3. A female in late pregnancy (near term) has the same energetic requirement as a nonpregnant, nonlactating female of the same total weight [*basal metabolic rate may be slightly higher but activity level is*

probably slightly lower], so at term or immediately at the birth of her infant, the percentage of time spent feeding, f_p, is given by

$$f_p = A(m + i_0)^{.75} \tag{2}$$

where i_0 is the weight of the infant at birth and m is the weight of maternal tissue. I further assume that maternal weight is constant from full-term pregnancy throughout the lactational period [*I am ignoring placental tissue and fluids*].

4. A mother's energetic needs while lactating are a result of energy required to maintain her own weight plus energy required to maintain her infant. The infant's weight, i, is a function of t, the infant's age. I shall assume that the infant's activity level remains constant throughout the first year of life and is the same as its mother's, that is, A is the same for both. [*In the first two months of life, infants' activity levels are probably lower than those of their mothers; for older infants activity levels are probably higher. Incorporating this change, a more realistic model would predict a lower percentage of time spent feeding for these early months but higher percentages after month two, compared with the percentages in the present model.*] I also assume that the infant requires energy only to maintain its weight at each age [*that is, for the time being, I am ignoring the energy required to produce new tissue*].

Then f_t, the percentage of time that a mother needs to spend feeding when her infant is age t, will be given by

$$f_t = Am^{.75} + \frac{A(i_0 + t\,\Delta i)^{.75}}{E} \tag{3}$$

where E = net efficiency of maternal lactation (ratio of milk calories to lactational increment in maintenance calories) and of assimilation by the infant; Δi = increment in infant weight per unit time, t, here kilograms per day.

We can now evaluate this function utilizing currently available estimates of the relevant parameters.

From laboratory data and trapping records I estimate $i_0 = 0.775$ kilogram (Snow 1967) and $m = 11.00$ kilograms (using Bramblett 1969 and Snow 1967). This was calculated on the following basis. Baboon birth weights in the laboratory and human birth weights are relatively independent of maternal weight and diet (Buss and Reed 1970). Thus it seems reasonable to assume that infants born in Amboseli are approximately the same weight, 0.775 kilograms as those born in laboratories (Snow 1967). Maternal weights in the field, by contrast, are probably at least 1 kilogram lighter than those in a laboratory. It is well known that adult weight is a function of diet and activity. Weights of wild-caught baboons average less than those of baboons in laboratory colonies (11.5 kilograms calculated from data in Bramblett 1969, versus 12.3 given in Snow 1967; however, Bramblett's data presumably include some pregnant females; thus my estimate of 11 kilograms for nonpregnant Amboseli females). Even in Amboseli, the baboon groups that live a few miles from the study group and frequent the garbage dumps of the park lodge all appear to be considerably larger than any animal of comparable age in the nonlodge groups.

From my field data (Fig. 12) I estimate that $f_p = 45$. From equation (2) we then have

$$A = \frac{f_p}{(m + i_0)^{.75}} = \frac{45}{(11.775)^{.75}} = 7.08$$

Then from equation (1) the percentage of the daytime required for feeding just to enable a mother to maintain her own body weight, 11 kilograms, is given by

$$f_m = Am^{.75} = 7.08(11.00)^{.75} = 42.76$$

Finally, I estimate $.005 < \Delta i < .010$ (Buss and Reed 1970, Snow 1967), I assume Δi constant over the first year of life (Snow 1967), and estimate $E = .80$ (Blackburn and Calloway 1976b). [*Brody's (1945) mean value of 60 percent for net efficiency of milk production would only exaggerate the consequences discussed below.*]

Finally,

$$f_t = 42.76 + \frac{A(.775 + t\ \Delta i)^{.75}}{E} = 42.76 + \frac{7.08}{.80}(.775 + t\ \Delta i)^{.75}$$

This function is graphed, producing the two sloping "lines" (literally curves, since age appears in the equation to the .75 power, not 1) in Fig. 14 for $\Delta i = .005$ and $.010$ kilogram/day. The values $\Delta i = .0052$ for the female and $\Delta i = .0067$ for the male are the two values obtained by Buss and Reed (1970) for two infants whose mothers were fed a low-protein diet. The mothers were being maintained on the lowest-protein diet at which Buss found they could maintain their weight. Buss and Reed discontinued their study of infants of mothers on low-protein diets after four months because the infants dropped below the range of normal values obtained in his laboratory. The values $\Delta i = .0098$ for males and $\Delta i = .0085$ for females are the values that Buss and Reed indicate are the normal growth rates in their laboratory.

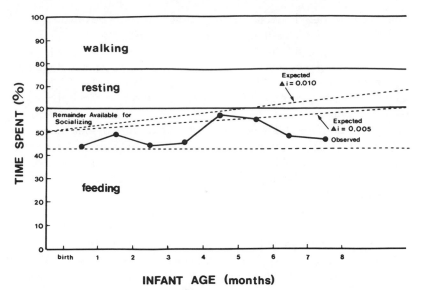

Fig. 14. *Predicted and observed amount of time mothers spent feeding, based on a model of necessary feeding time for mothers who provide all their infants' energetic requirements (see text for details).*

A value of .0082 for Δi of both males and females was obtained by Snow (1967) in a laboratory study of growth and development of approximately eight baboon infants.

Tabulating various landmarks of physical development through ages of reproductive maturation as reported by Snow and as obtained in the field in Amboseli, I found that the Amboseli growth rate compares with the laboratory rate as about 3 : 4.5. Applying this to obtain an estimate of Δi in Amboseli gives a value for Δi of 5 to 6 grams per day, .005 $< \Delta i <$ .006, corresponding closely to the values obtained by Buss and Reed for infants whose mothers were on low-protein diets.

The value of 43 percent of the daytime spent feeding needed for a mother to maintain her own body weight is calculated from time budgets in which females spent approximately 17 percent of their time resting and 23 percent of the time walking—a total of approximately 83 percent of the time spent in these three activities, as indicated by the three horizontal lines blocking off sections of the time budget graph in Fig. 14; one line for feeding at 43 percent, blocking off a section from zero to 43 percent, one at 77 percent marking off an area from 77 percent to 100 percent indicating the time spent walking; another at 60 percent blocking off a section from 60 percent to 77 percent for the 17 percent of the day spent resting. The remaining time might be considered "uncommitted," time that could be devoted to so-cializing, or to more resting, feeding, or walking.

According to the model, at parturition a female would have to spend another 7 percent of her time (over her own 43 percent mainte-nance level) feeding just to provide energy for her newborn infant, re-sulting in the intercept of the two sloping "lines" on the graph. Then the other values on these two lines indicate the additional time that a mother would have to devote to feeding at two rates of infant growth if all the additional energy she obtained went to the infant and not to the mother herself (except for energy she used for the infant). The rates of infant growth probably represent the extreme ranges that might occur, as indicated above.

It is clear from Fig. 14 that a female would have to increase her feeding time from the 43 percent of her time spent feeding to maintain only her own body weight to 58–66 percent during month nine of in-fant life, if she were providing all of her infant's energetic needs and were doing so only through lactation (these percentage incremental values are consistent with those found for humans; see Whichelow 1976). She would still need to spend approximately 23 percent of her time walking for a total of 81 to 89 percent of her day occupied just by feeding and walking. However, at these levels she would no longer be able to rest for 17 percent of the day, even if she could totally eliminate time spent socializing. The result is that her overall activity level would in fact be greater, creating even greater energetic requirements for her own maintenance, and she would therefore need to spend even more time foraging.

The conclusion we must reach is that even with fairly conserva-tive estimates of energetic demands, a mother could not provide all caloric requirements for herself and her infant beyond six to eight months of infant age and probably could do so up to that age only with difficulty and major restructuring of other aspects of her life. If the model were modified to incorporate the refinements I have indicated, would the apparent required feeding time be reduced? Most assump-tions were conservative and refinements would lead to more intense time budget constraints, especially after the infant's first few months. Perhaps, then, mothers and infants have had to accommodate to these constraints as a reality. Perhaps mothers cannot always maintain their own weight and infants must provide some of their nutrition. If so, in-fants' maturation and learning and other factors that facilitate the tran-sition to nutritional independence have probably been under consider-able selective pressure. It is useful to turn to the observed feeding time budget for the mothers of the present study, which is also plotted in

Fig. 14. If we examine the observed values obtained during this study, it is clear from Fig. 14 that a female's time budget is affected by the fact that she has a dependent infant and by the age of that infant, but that the effect of infant age is less than that which was predicted above. Mothers are probably not maintaining their body weight (see Hytten and Leitch 1963, Naismith and Ritchie 1975 for humans). It is known that women who lactate successfully and whose caloric intake during lactation is 23 percent over their normal intake do not maintain their body weight, whereas at 32 percent, women maintain steady body weight (Whichelow 1976). If such weight losses occur in baboons, insufficient nutrition and weight loss may be a major source of maternal susceptibility to death and may place severe limits on the length and intensity of the lactational period unless females are able to store an appreciable supply of excess fat during pregnancy. Because reduced maternal health would directly affect the chances of survival of the mother's current infant, there will be immediate and evolutionary pressure for factors that enable infants to provide some of their own energetic requirements.

The time course of energetic demands seems to be as follows. During the first two months of infant life the additional amount of food that a mother requires to support a dependent infant over the amount required during pregnancy is primarily due to the lower efficiency of lactation but is also due to the energetic requirements of retrieving and attending to the infant when it is out of contact and to a small amount of infant growth. In the next three months the infant's continued growth and increased activity level place considerable strain on the mother's ability to maintain her own weight because the infant cannot contribute appreciably to its own nutrition at this stage; but also I suspect that mothers lose weight during this period. If milk supplies are reduced owing to maternal nutritional strain and weight loss, infants may have additional "motivation" to eat the many plant foods that they explore. By the time their infants are five or six months old, mothers are feeding all the time they can (60 percent) without sacrificing a considerable portion of social time and/or rest time. If they sacrifice rest time or time spent being groomed in favor of other activities, they will need to feed even more because any other activity would require more energy than these do. Thus it appears that mothers may have reached a maximum of feeding time by the time their infants are five to six months old, perhaps even with weight loss. It is surely necessary for their infants to provide considerably for some of their own nutritional needs. The infants can more readily do this at five or six months of age if the right foods are available.

Weaning Foods

Recently, anthropologists have emphasized the importance of availability of so-called weaning foods as a major determinant of age of weaning, and consequently of the amount of lactational energetic demand on the mother and of the length of postpartum amenorrhea and interbirth intervals (see Howell 1979, Lee 1978). What is a weaning food? For a human infant, it is a food that is both easily eaten (one that is soft and smooth, like porridge) and readily digested. For a baboon, whose mother neither collects nor prepares its food, the ease of obtaining the food must also rank among the major criteria of a weaning food.

A detailed analysis of feeding behavior and nutrition of weanling baboons in Amboseli is under way (S. Altmann in preparation), as is a related study of vervet monkeys (*Cercopithecus aethiops*) in the same habitat (D. Klein in preparation). However, the extremes of food accessibilities are quite obvious even to the casual observer. Flowers, particularly those of umbrella trees (*Acacia tortilis*) and fever trees (*A. xanthophloea*), are abundant seasonally in Amboseli and easy to pluck. Any infant who can negotiate within the trees (i.e., any infant older than a few months) can readily eat large quantities of these sweet tid-

Fig. 15. *Infants of various ages exploring plants at the edge of a waterhole. The youngest, three-month-old Oreo, is at front, right.*

bits. In at least some years, umbrella trees flower more heavily than do fever trees (Post 1978 and personal observation). Their branches are more horizontal and have rougher bark, and therefore can be negotiated more readily by infants and at a younger age than can fever trees. Many mature blossoms come loose and fall to the ground as the baboons feed. Thus even Pedro, who could not climb at all and had no fine motor coordination in his hands (see Appendix 2), was able to eat umbrella tree blossoms, picked from the gound, as virtually his only food other than milk. It may not be coincidental that he lived two months (late December to early March) after his impairment became severe, when we thought that he would die any day. He died at the beginning of March, soon after the period of abundant acacia blossoms was over.

Fruits and berries (e.g., *Tribulus terrestris, Azima tetracantha, Salvadora persica*) and the freshest green grass blades and leaves are probably the next best weaning foods; however, skill is particularly required to pluck *A. tetracantha* berries without being stuck by the abundant sharp thorns of this shrub and small infants can only reach the lowest branches (S. Altmann, personal communication). Gum of very small, shrublike fever trees is probably also a good weaning food—it is high in carbohydrates (Hausfater and Bearce 1976)—but to obtain gum from the larger fever trees, such as are used by the baboons as nighttime roosts, requires an ability to negotiate the long, smooth, upright trunks and branches of these trees. The youngest green acacia pods are probably fairly accessible and manageable, but the youngsters apparently need to extract the seeds and discard the pods themselves, although adults (at least the males) eat virtually the whole pod (S. Altmann, personal communication). Consumption of the acacia seeds probably is limited by the presence of toxins, to which the young animals may have a lower tolerance than do adults (S. Altmann in preparation).

At the other extreme are grass corms, one of the few foods available by the end of the dry season (October), the harvesting and consumption of which account for the largest proportion of the adult baboons' time budget except in January, February, and August (feeding on berries and gum occupies almost as much time in December) (Post 1978). Even older infants cannot dig anything other than the smallest corms. This is probably the major food that infants scavenge, sorting through the discarded scraps at a place where their mother or a male associate has been digging.

Is there a season when these weaning foods are most available, and is this time fairly circumscribed? For 1974–75, some relevant data are available from Post (1978). (For 1975–76 much comparable data will be available in S. Altmann in preparation and in D. Klein in preparation, but the timing of changes in the tree phenology in particular seems to be fairly invariant from year to year despite annual variations in rainfall patterns.) During 1974–75 over 50 percent of the umbrella trees had blossoms in December and January; about 15 percent in November, February, and March; none in other months. In December, January, and February over 10 percent of the trees had heavy blossom production, more than in any other months. During no month were as many as 10 percent of the fever trees in heavy flowering, but in October and November about 20 percent of the trees had some flowers, and in December about 5 percent. Ripe *Azima* berries were most abundant in January, and were very accessible in February and March.

Thus available data on plant phenologies indicate that December, January, and perhaps February are the months in which an infant could most readily start taking solid foods. From observations on development of both gross motor coordination for climbing in the trees and fine coordination for manipulating foods, I would estimate that a baboon infant in Amboseli would have to be at least three to four months of age even to begin to utilize much weaning food, and perhaps another two months older before attaining sufficient competence to contribute appreciably to its own nutrition. In the previous section it was also seen that mothers probably find it increasingly difficult to provide infants of this age with all their energetic needs. Infants that are conceived in December and January are born in June and July and would be at the right age to utilize weaning foods when these are most available. Therefore, one would predict the highest survival rates for infants born in June and July. Moreover, because food is apparently also most available to adults in December and January (these being the months during which the least time was spent feeding, during both 1974–75 and 1975–76), it is also reasonable to expect highest conception rates during these months.

Thus both factors suggest selection favoring a birth peak in June and July. I examined the seasonality of conception and the differential survival of infants born at different times of the year. The suggestion that food availability affects baboon reproduction in Amboseli is substantiated by the available data. Of 54 pregnancies for which we know the month of conception, 13, or 24 percent, of the conceptions occurred in the two months December and January, a result in the predicted direction but not statistically significant ($P < .12$, one-tailed binomial with probability $= .17$, or one-sixth). However, conception by females with a surviving semi-independent offspring would be expected to be even more subject than other females to the effects of the seasonal availability of food. Of 34 such conceptions, 11, or 32 percent, occurred during December and January, about twice the number one would expect without seasonality ($P < .03$, binomial test as above). What of the survival chances of infants born at different times? The infants conceived in December and January also have higher chances of live birth and first-year survival ($11/14 = .79$) than do infants born at other times of the year ($20/36 = .56$). (The sample size is slightly smaller for this comparison than for the previous one owing to incomplete survival data for recent births.) Despite the large difference in these two values, sample sizes are not adequate to reach statistical significance.

Given the apparent advantages of seasonality, one might ask why infants are born any time other than June and July, and why mothers conceive any time other than December and January. Such extreme seasonality would entail interbirth intervals of either one year or two. That baboons are capable of one-year intervals is clear from the fact that these are the norm in some zoos (e.g., Lincoln Park, in Chicago); Thelma Rowell (1966b) gave the period of postpartum amenorrhea as five to six months and a subsequent cycling time as one to three cycles in her caged colony in Kampala, Uganda, and perhaps also in the rich riverine habitat of Ishasha, Uganda. Elsewhere, however, in the widely varied habitats of Gombe, Gilgil, and Amboseli, postpartum amennorhea is of the order of ten to twelve months if an infant survives (J. Altmann et al. 1977, Packer 1979a, Ransom and Rowell 1972, Sigg and Kummer in preparation). A subsequent cycling time of about four to five months, then, does result in almost a two-year interval.

There are two situations in which one would expect "off-season" conceptions and infant survival to be higher than usual. The first is an occasional extended rainy season or other factor that leads to an extension of the season of good weaning foods and of generally higher food availability. When this occurs, females will be more likely to conceive during these off-months and infants who were born at nonoptimal times and are semi-independent when this extended good time occurs will be more likely to survive. Occasional good years may reduce the ecological pressures for birth seasonality. The second situation occurs when a female's previous infant dies. Then the female is faced with a "decision"—conceive now or wait perhaps several months or more for the best conception time. If she has no dependent infant, she will probably be in better physical condition and therefore more able to conceive and to produce a healthy infant with good survival chances. Also, by doing so she will not be endangering the survival of a present young offspring because she has none. Thus the variability in advantages of conception during different seasons is not as great for these females and they are more likely to have off-season conceptions. Since high infant mortality is characteristic of most nonexpanding populations, it will often tend to reduce seasonality of births in a habitat and species for which factors favoring birth seasonality are not already very strong.

These two situations probably account for the absence in the literature of a clear demonstration of birth seasonality for any baboon population (but see Lancaster's and Lee's 1965 summary of the available evidence as of that date and Keiding's 1977 reanalysis of our 1963–64 Amboseli data). Kummer (1968) described two birth peaks in the hamadryas baboons but also noted that birth peaks occurred at different times in harems in the same area. More recently, Abegglen (1976) found a late spring birth peak in several bands of hamadryas over a period of several years. In Amboseli, the 1963–64 data and our less complete records on other groups since 1971 all indicate a seasonality similar to that documented here for Alto's Group.

Summary

Seasonal variability in rainfall and food availability is reflected in the baboons' daily time budgets, the average distance they travel each day, and in the seasonality of births and infant survival. A model of a mother's feeding time, based on a mother's providing all her infant's nutrition, resulted in a prediction of severe time budget problems for mothers of older infants even at quite slow rates of infant growth. Observed feeding time was below predicted amounts for these mothers. I suggest that the compensating factors are (1) maternal weight loss to compensate for energetic requirements that cannot be met within a tolerable time budget and (2) eventually infants' providing some of their own food.

It is important to note that no category labeled "infant care" was included in mothers' time budgets. For baboon mothers as for human mothers, most infant care is done concurrently with other activities. This is true even when the infant is not in contact; mothers are probably "tuned" to their infants, alert to possible trouble. We do not yet know how infants affect mothers' abilities to attend to other stimuli or how efficiency of other activities is affected by concurrent infant care for any primate. It seems likely that the efficiency of activities such as feeding is lower than it would otherwise be for females when they have infants and that this might result in further strain on mothers.

Reproductive success and survival of Amboseli mothers and infants were seen to be subject to considerable environmental influences. It is not unreasonable to expect that their behavior and physiology have been molded through natural selection to respond to these conditions and that they also respond facultatively (flexibly) to make the best use of alternatives in the particular situations in which they find themselves at any given time. In the next chapters I shall consider social and other behavioral interactions, between mothers and infants and between each of them and other group members; the developmental course of these interactions; the meshing of these behaviors with the less social aspects of their world that have been considered thus far; and the extent to which they relieve or intensify pressures on mothers and infants.*

* Births from 1 August 1978 through 31 December 1979 exhibit less seasonality than do those reported for the previous years (pp. 60–62), perhaps owing to two years of unusually high rainfall and its effect on plant growth. It is too early to determine the first-year survival rates for these infants.

6 / Social Milieu

JUST AS THE MONTH of birth and other aspects of seasonality may result in individual differences in the ecological pressures on mothers, so may individual differences in social experiences affect the mother and her infant, both directly and indirectly through effects on the mother. What are these social experiences? How do they affect mothers' time budgets and create attentional demands, and in what other ways do they appear to improve or to reduce maternal and infant survival? An understanding of the sources of these differences is the first step toward determining potential lifetime and intergenerational continuity in such experiences. At parturition the social life of a baboon female changes dramatically. She must not only nurse, carry, and protect the neonate while still providing her own food, transport, and protection, as before; but in addition, she and her infant become a major focus of interest within the group (DeVore 1963, Ransom and Rowell 1972, Seyfarth 1976), a common characteristic of primates (see Hrdy 1976).

Several field studies have resulted in qualitative or normative descriptions of the social world of savannah baboon mothers and their infants: those of DeVore (1963) and Rowell and Ransom (Ransom and Ransom 1971, Ransom and Rowell 1972), all for olive baboons, *Papio anubis*. In addition, some quantitative data are provided by Nash (1978) primarily for 6-to18-month-old anubis baboons and their mothers. Cheney (1978) and Seyfarth (1976) present some relevant quantitative data from a general field study on a small group of chacma baboons, *P. ursinus*. Rowell et al. (1968) studied mothers and infants in a small captive colony of anubis baboons.

Most general descriptions of the social relations of baboon mothers are consistent across studies. In particular, all authors report

Fig. 16. *Several juveniles crowd around Slinky and her new infant, Sesame.*

that mothers of young infants receive little overt aggression, more grooming, and more approaches than do females in other reproductive stages. This general pattern of interactions within a group has also been described from field and laboratory studies of most other primates (see Hrdy 1976, Lancaster 1971) and probably comes closer than any other to being a primate "universal." One might be tempted to assume that the social group provides support for and reduction of the stress of motherhood. However, as I discuss below, a baboon mother's social world may not be as benign as these descriptions imply, and occasional extreme violence toward primate mothers or infants provides rare but chilling exceptions in some species (see, e.g., Angst and Thommen 1977; Goodall 1977; Hrdy 1974, 1977; Sugiyama 1967).

The social group has an impact in both active and passive ways. The active ways are fairly obvious: individuals groom each other, play, fight, and so on. Most of the time, however, individuals, even close neighbors, are not interacting. But the mere presence of conspecifics can have a social effect even when these animals are not engaging in a social interaction. The presence of individuals who are alert to danger or who locate possible food sources may be beneficial to nearby group members. Spatial arrangements within groups also provide the broad outlines for social opportunities that may be used or not—a large adult male, just by his presence, may discourage others from approaching a mother; the mere presence of an individual who has harmed a mother previously may disrupt her activities or result in her restricting her young infant's movements. Thus consistent spatial patterning can be a clue to rarely observed interactions. In the sections that follow, before analyzing the more obviously beneficial or harmful interactions, I shall first examine the spatial patterning, that is, neighbor relationships. Then I shall turn to approaches: the interface between the statics of spatial relationships and the dynamics of interaction. Finally, I shall examine two major forms of social interactions, apparently beneficial social grooming and stressful agonistic interactions, including spatial displacements, or supplantations.

Sleeping Grove Subgroups

Most of our observations of baboons are made during the daytime. At night the animals are high in the sleeping trees. Their location and the lack of available light make observations most difficult. Fortunately for us, it appears that little activity usually goes on at night; this is suggested by two all-night watches that we have made and by the fact that individual animals can usually be relocated in the morning in the same place that they were seen to settle into the night before. Yet spatial relationships at night may be quite important. Early-morning grooming sessions often occur in the trees between individuals who sleep near each other. Also, predation seems to occur primarily at night, a fact that perhaps renders nighttime spatial relationships particularly important; for example, smaller animals may be less subject to predation if they sleep next to an adult male. Then, too, close neighbors may provide each other with warmth during the cool, dry savannah nights. Finally, the nighttime relationships, by their static nature, may elucidate patterns that exist during the day but are overlayed with foraging and with general activity levels that produce a more fluid social world. Therefore, it seems appropriate to consider first the sleeping grove subgroupings that constitute the spatial milieu during 13 to 14 hours each day.

During 1975–76, although Alto's Group used 15 to 20 groves of trees that were scattered throughout the group's home range, one was used most often, over 40 percent of the time. This "Favorite Grove," as we dubbed it, consisted of two trees with continuous crown, and a third tree that was sufficiently separate that animals moving from one subgrove to the other went down one trunk and up the other rather than attempt the long jump across the branches.

It appeared that the same individuals consistently used each of these two parts of the grove: in fact, individuals were very consistent in where and near whom they slept even within their respective trees. Although I could not keep consistent records of the latter, I did record which tree subgroup (labeled "left" and "right" in my records and in Fig. 17) each group member was in whenever the group slept in Favorite Grove and members were still settled in their sleeping postures when I arrived in the morning. All members were so located and identified on each such sample day unless, as occasionally happened, descent occurred too rapidly or too soon after my arrival. This sampling was begun in February 1976, when I could identify with confidence all individuals in virtually any position while they were huddled high in the tree, something that I could not do in the early months. I have data for 44 days that the group used Favorite Grove, from February through July and in October. For those baboons who neither died nor migrated during this period I have data from about 35 to 40 days in which subgrouping occurred (none occurred on 2 nights).

Ad libitum observations of spatial patterning in other sleeping groves were consistent with those in Favorite Grove, especially at the level of the three or four closest animals in small subgroups. However, comparable quantitative data could not be gathered for these groves

"LEFT GROUP"　　　　　　　　　　　　　"RIGHT GROUP"

Adult and Subadult Males

Percentage of Samples				Percentage of Samples		
All	> 90%	51-90%		51-90%	> 90%	All
	♂ Slim				♂ Even	
	♂ Peter				♂ Max	♂ Ben
♂ Stiff				♂ High Tail		
				♂ B J		
				♂ Chip		
						♂ Red

- - - ♂ Russ - - ->

<- - - ♂ Stu - - -

Juveniles

♂ Swat
♂ Nog
　　　♂ Dogo
　　　　　♂ Toto
　　　♀ Janet
　　　♀ Dotty
　　　♀ Nazu

- - - ♀ Cete - - ->　　　　♂ Jake

<- - - Striper - - -

Adult Females and Infants

　　　　　♀ Plum (♀ Peach)　　(♀ Pooh)

♀ Oval (♂ Ozzie, ♀Oreo)　　　　　　　♀ Fem (♂ Fred)
♀ Alto (♀ Alice)　　　　　　　　　　♀ Este (♀ Eno)
♀ Scar (♀ Summer)　　　　　　　　　　　♀ Brush (♂ Bristle)
♀ Preg (♂ Pedro)　　　　　　　　　　　♀ Handle (♂ Hans)
♀ Judy (♂ Juma)　　　　　　　　　　　♀ Gin (♂ Grendel)
♀ Spot (♀ Safi)　　　　　　　　　　♀ Slinky (♀ Sesame)
　　　♀ Vee (♀ Vicki)　　　　　　　　　♀ Lulu
　　　　　　　　　　　　　　　　　　♀ Mom (♂ Moshi)

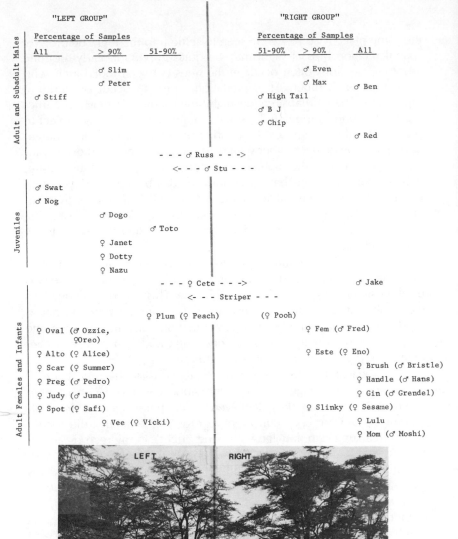

LEFT　　RIGHT

owing to a lack of clear physical division within the groves, infrequency of grove use, or poorer visibility within some of the groves.

The subgroup data yield information about the affiliative patterns within age classes and the maturational trends in relationships of mothers, infants, maternal siblings, and cohorts.

Almost all individuals were at least 90 percent consistent (no more than 4 switches out of 40 samples) in their subgrove choice (Fig.17), most of them making no switches at all despite births of infants and other changes in reproductive state. Of the fully adult baboons, only male High Tail with 5 switches in 13 samples taken before his death and female Plum with 12 switches in 33 were less consistent. Adults consistently in the left subgroup were males Slim and Peter and females Alto, Spot, Vee, Preg, Scar, Oval, and Judy. Consistently in the right subgroup were females Mom, Lulu, Fem, Gin, Slinky, Handle, Este, and Brush and males Even, Ben, and Max.

Of the young adult males, Russ and Stu gradually made the transition from one subgrove in February to the other by October, moving in opposite directions. Among the older juveniles and subadults, Toto made 7 switches in 37 samples. The two- to three-year-olds tended to sleep in the same subgrove as their mothers but not uniformly so (see Striper and Cete, who gradually changed subgroups during these months).

All animals less than two years old slept in the same subgrove as their mothers except Pooh, who had severe locomotor impairment (see Appendix 2). She slept with or near adult Max in the rightmost tree, which she could climb more easily than the others. Fred slept with his mother; Alice did so until Alto's death, at which time Alice began to sleep with adult male Peter or her sister Dotty (Alice was 17 months old at the time). Ozzie slept with his mother, Oval, until the latter part of Oval's next pregnancy, when she overtly rejected him; he was then about 18 to 19 months of age. After that he sometimes slept with his older sister, Fanny.

Fig. 17. *Subgroup membership in Favorite Grove from February through October 1976. On the picture of Favorite Grove the sleeping trees are indicated with their subgrove labels. Each individual is listed with the group where it spent more of its time, except for those in the center section, where dotted arrows indicate a gradual unidirectional shift from one grove to the other. Within each subgrove, the individual's column placement indicates the percentage of time it spent in that subgrove. Individuals less than two years old are indicated by parentheses and their column placement should be read as that of their mother, except for Pooh (see text). During October 1976, the month of Peach's birth, Plum spent all nights in the right subgrove.*

Several factors probably led to the average relatedness within sleeping subgroups being higher than the relatedness in the group as a whole. As noted above, both infants and their next older siblings tended to sleep in the same subgroup as their mothers. Moreover, the adult males and females of a subgroup tended to be each other's preferred mating partners. Therefore, the fathers of a subgroup's infants were likely to be the males of that subgroup. Finally, all adult females (with the partial exception of Plum) who had been members of High Tail's Group before the merger, were in the same (right) sleeping subgroup. As we shall see below, a mother's sleeping subgroup tended to include the individuals who were her daytime neighbors and interactants.

Daytime Neighbors

During the daylight hours, the baboons spend most of their time foraging on the ground, where they move more easily than in the trees, sometimes gather in clusters, and are not restricted by particular comfortable locations, as they seem to be in the trees. Infants can more readily move about and do so at a younger age. In the daytime I studied neighbor relationships in finer detail. Note that when I examine the identities of mothers' neighbors and interactants, I usually focus only on the infant's first three months of life rather than older ages. This is done for two reasons: because the mother's and infant's worlds most completely overlap during these months and because, as can be seen below, these are the months of greatest social involvement. These are also the only months during which baboon infants possess their distinctive black natal coat. Coat change begins during month three and is mostly complete by the end of month six (see Chapter 8).

Two aspects of neighbor relations may be of considerable importance to mothers and infants: the density of neighbors and the identity of these neighbors. From W. D. Hamilton's work on adaptive group geometry (1971) and W. C. Allee's early studies of the advantages of sociality (1931), one would predict that being surrounded by more conspecifics for more of the time would be advantageous to an individual in that it would provide additional protection against predators. This would be even more true if, as is sometimes the case in baboons, this circle includes at least one adult male: the large canine teeth of adult male baboons are formidable weapons. Alternatively, other arguments (see, e.g., Alexander 1974) emphasize the potential disadvantages of being close to conspecifics, in particular the increased likelihood of disease transmission and of feeding competition. I shall

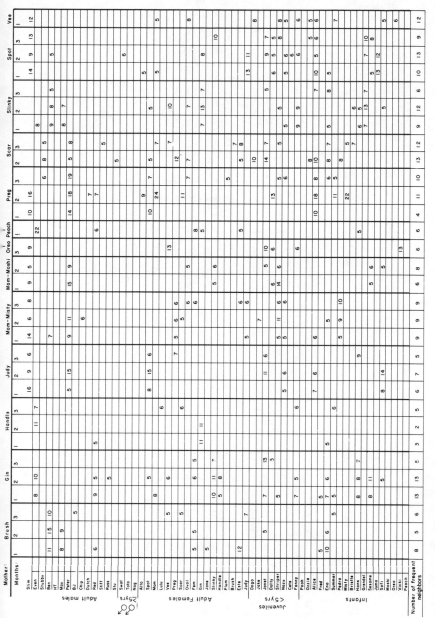

Fig. 18. Matrix of the percentage of time an individual was within 2 meters of each mother during her infant's first three months of life during which I obtained samples. Percentages are included only for those individuals who spent at least 5 percent of the time near the mother. The last row indicates the total number of these frequent neighbors for each mother.

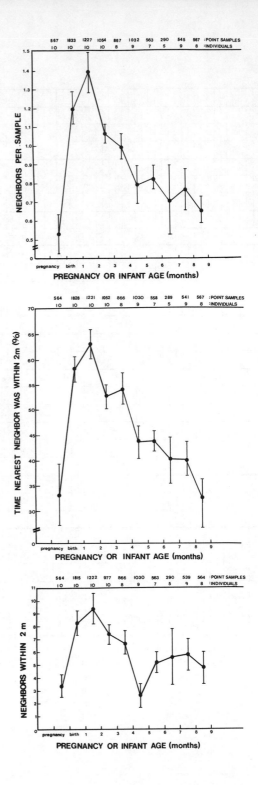

consider the relevance of each of these suggestions in the context of the neighbor patterns found in this study.

The point-sample data allow estimation of the percentage of time that animals spend in proximity to each other. I recorded the identities of all individuals who were within 2 meters of the focal female, a distance chosen because it seemed to represent an individual's "personal space": there were few animals that an individual would tolerate that close to it for long. Neighbor data for greater distances were also gathered but will not be discussed further here.

Only rarely did baboons spend 10 percent or more of their time within 2 meters of new mothers (Fig. 18). In almost every such case the individuals near a mother were either family members, one or two adult males, or another mother with a same-aged infant. Five percent of the time spent in proximity, defined as within 2 meters, was about the lowest level that I could reasonably expect to measure when partitioning each female's point-sample data by months. For each female and each month of infant life, all individuals who spent at least 5 percent of the time nearby were considererd.

Two ways of looking at neighbor densities provide similar pictures. For each month of infant life, the average number of animals within 2 meters, per sample, is depicted in Fig. 19, top; the average time that the pair had at least one other animal within 2 meters, estimated as percentage of samples in which the nearest neighbor was within 2 meters, can be seen in the middle panel of the same figure. These both show the same sharp increase in neighbor density during month one of infant life (i.e., between birth and one month of age) over that seen during pregnancy, some increase again in month two, and a consistent decrease after month two, dropping to late pregnancy levels by about month seven to nine. Note that the average number of neighbors per sample increases by approximately the amount that we would expect from the increase in the amount of time that there is at least one neighbor within 2 meters. That is, new mothers are "alone" about half as often as when they were pregnant; however, when there is at least one animal nearby, the number of neighbors is approximately two dur-

---

Fig. 19. *Neighbors within 2 meters during the last month of pregnancy and at each month of infant age. The top graph depicts the average number of neighbors per point sample, i.e., the average number of close neighbors an individual had; the middle graph shows the average percentage of time that the nearest neighbor was within 2 meters, i.e., that there was at least some other baboon close by; the bottom one depicts the average number of different individuals who spent at least 5 percent of the time nearby. See also Fig. 18.*

ing pregnancy and at each infant age. We do not yet know the extent of protection against predation that is afforded by this pattern of neighbor spacing, but some degree of protection seems reasonable. This type of increase in neighbors—increased time with at least one neighbor but not high densities of neighbors—probably does not result in appreciable increases in feeding competition.

The opportunities for disease transmission may well be greater than during pregnancy because they probably depend not only on the time spent with others nearby, but on the number of different animals to whom a mother and infant are exposed. Thus, examining the bottom panel of Fig. 19, we see that during the last month of pregnancy, females averaged three or four different individuals who were their frequent neighbors (median two or three). Although after parturition the set of frequent neighbors still represented a small fraction of the total group membership of about 45 animals, for all of the females the number of different neighbors was greater during the first month of infant life than during the previous month, for an average of eight different individuals (median 8 to 9) who spent more than 5 percent of the time nearby. Nine individuals spent at least 5 percent of the time nearby during month two, six or seven in month three, five in month four. Thus, many more different individuals were spending at least 5 percent of the time near new mothers than had done so near pregnant females, probably increasing the likelihood of disease transmission.

Both feeding competition and likelihood of disease transmission may also depend on variability in the tendency of particular individuals to engage in harmful or helpful activities when they are nearby. Some individuals may provide more assistance or threat than others, both immediately and as potential long-term associates. Who were the individuals who were frequent neighbors?

The tendency of particular adult males to associate with particular mothers is clear from Fig. 18. For example, both Brush and Mom with Moshi had two adult males who spent time nearby, but it was Ben and Max for Brush, Peter and Slim for Mom. After an infant was about two months of age, some associated adult males also spent time with the infant when it was separated from its mother, a factor not reflected by these data, which are only on mothers. Male associates were members of the mother's sleeping subgroup and usually had previous associations with the mother, sometimes including mating when the infant was conceived. Their interactions will be examined in more detail in later sections.

Although four- to six-year-olds interacted with active, especially semi-independent, infants, they rarely spent much time near the mothers. During the year of this study, all members of this age class were males (Dogo, Nog, Toto, and Swat). Females of this age class are approaching menarche and their first year of cycling (J. Altmann et al. 1977, Packer 1979a). Their interactions with mothers and infants might be quite different and are now under study in Amboseli (Stein in preparation, Walters in preparation).

The two- to three-year-old juveniles varied considerably in the time they spent with new mothers, even with their own mothers when they had infants (Fig. 18). Females Cete and Nazu were only a year-and-a-half old when their mothers gave birth to Summer and Pedro; it was during those pregnancies that these young juveniles were probably weaned. They did not stay near their mothers during the first months after Summer and Pedro were born and only later interacted with their siblings. The same was true for male Ozzie and female Pooh when their mothers' next infants were born (see below). Females Janet, Striper, and Fanny, by contrast, were two- to three-year-olds when their current siblings were born, and each of their mothers had experienced at least one other pregnancy since their births. They all showed considerable interest in their siblings, as did two-and-a-half-year-old female Dotty when her sister Spot gave birth.

With family data excluded, there are still clear overall differences among these juveniles. Janet, Nazu, Fanny, and Striper were common neighbors of mothers: each of these juveniles was included in one-fourth to one-third of the data for mothers with infants in the first three months. In contrast, Dotty, Cete, and male Jake appeared in less than one-tenth of the data. Thus most juveniles who tended to be interested in infants to whom they were related were interested in most unrelated infants as well. Neither age nor dominance rank explains the differences among the juveniles. Follow-up studies of these juveniles (Walters in preparation) and of the later cohorts are under way.

The yearlings spent much less time near new mothers than did the two- to three-year-olds, and the time they did spend was usually spent in conjunction with their own mothers. Often a weanling would get on its mother's nipple while its mother groomed a new mother. Female Alice, offspring of the highest-ranking female, was an exception: she was highly interactive in general and spent time with new mothers, often sitting in their ventrums and pushing aside these mothers' infants. She spent much time near Judy, Vee, and her sister Spot. Male Ozzie also spent considerable time with Spot and Vee but not with his mother, Oval, when she had a young infant. Female Eno and male Fred often were near Gin. Female Pooh was virtually noninteractive and spent little or no time with any of the new mothers, including her own when Peach was born.

No general patterns emerged from an analysis of adult female neighbors of mothers except that Gin and Mom were particularly likely to be neighbors of other mothers during the first month of life of the other mothers' infants. Some pairs of females who gave birth within a week spent much time near each other (Spot and Judy, Gin and Slinky); others did not (Mom and Brush, Vee and Plum). But even for those pairs that spent time near each other, the maintenance of the spatial relationship was often quite one-sided, as is shown in a later section. Moreover, some individuals who spent time together interacted seldom, and some frequent interactants did not spend much time together.

It is also clear from Fig. 18 that mothers varied considerably in the number of frequent neighbors they had, particularly during the first two months. Gin, Vee, Spot, and Scar had the most neighbors; Brush, Handle, Plum, Judy, and Mom with Moshi (but not with Misty) the fewest. In general, the former group is composed of females who are younger, rank higher, and have female infants, but it is not clear what variables determine variability in neighbor densities for neonates and their mothers. Females whose infants were born at about the same time varied—Vee with many neighbors and Plum with few, Spot with many and Judy with few, Gin with more than Slinky, Scar with more than Preg, Mom (with Misty) with more than Brush. Looking at these matched pairs does not clarify the picture. The direction is consistent with rank for four of five of these pairs, with age for four of five, and with infant gender for three of four (Plum and Vee both had female infants). Mom would be considered old and high-ranking when she had each of her infants. She had more close neighbors with her female infant, Misty, than with her male infant, Moshi.

Whatever the cause, some infants more than others grow up in frequent, close proximity to many individuals. This may speed their social maturation. It may also increase their exposure to disease and increase the amount of feeding competition that their mothers encounter, but there was no clear relationship between number of frequent neighbors and disease or mortality in this study.

In summary, mothers spent more time than others in close company with at least one other animal, and more individuals spent time near new mothers than near females without infants or mothers with older infants. Particular members of each age-sex class tended to spend time near mothers and infants. Some individuals, particularly adult males, spent considerable time with some mothers, and no time with others. Some mother-infant dyads had many individuals frequently in close proximity; others had few. Kin relatedness was not a major predictor of observed individual differences.

I have suggested some implications of the results for feeding competition, disease transmission, and predator protection. To further evaluate neighbors' potential effects, it is necessary to examine the nature of the interactions that occur among individuals. I shall turn now to some of the dynamics of relationships, first the tendency of individuals to seek each other out, that is, to approach each other, and then agonistic and grooming relationships. In all the analyses of this chapter, interactions that were only between a mother and her own infant are excluded. These are treated separately in Chapters 7 and 8.

Approaches

Social bonds are often measured by considering only active behaviors, such as grooming or sexual mounting for evaluating positive bonds, or aggressive encounters for evaluating repulsive relationships. Yet it seems reasonable that calm acceptance and shared proximity are also important. Such relationships, if they exist, contribute to the spatial patterns seen and described above, and provide the physical availability of animals for more active behaviors such as grooming, play, infant carrying, and perhaps even fighting; but it seems likely that they more often represent positive or helpful relationships than antagonistic ones. Therefore, I shall first consider close approaches, accompanied by no other social behavior and response, that is, those that result in the two animals just sitting or standing in proximity. Then in the next section I examine approaches that are part of a sequence of social interaction. Spatial displacements, or supplantations, where it is the place rather than the individual that is approached, are considered separately in the section on agonistic interactions.

SIMPLE, OR NONINTERACTIVE, APPROACHES

For practical reasons, approaches by one individual to another usually can only be identified, in the absence of a response by the second individual, when the approacher comes within a few meters of its object. I recorded as simple, or noninteractive, approaches all those approaches that carried the actor at least 1 meter (approximately a body length) and that brought the actor to within 2 meters of the animal being approached. These could be reliably scored without response by the object animal. As noted above, 2 meters seemed to represent an animal's personal space. Thus I wanted to determine which animals would be approached that closely with no further interaction and which animals' approaches to such proximity would be tolerated by an animal who was approached. With which animals do such calm approaches occur? Are they more common or less after parturition?

Fig. 20 indicates the rate of approaches to or by females for all individuals (cells) with values of at least .2 per 100 minutes, and in the last row the total rate for each female with all individuals. Those few individuals for whom rates met the criterion accounted for most approaches involving these mothers. For all 13 females the average rate of approaches to them was greater than the approaches by them (Fig. 20, last row). Each mother had a small set of individuals whom she approached, and this was to a considerable extent a subset of those from whom she tolerated approaches. In particular, most mothers had one adult male with whom they shared these approaches, for example, Mom-Moshi with Slim, and Brush with Ben.

There were usually more individuals who approached mothers than the mothers themselves approached. From the last month of pregnancy to the first month of infant life, both of these total rates (to and by the mother) rose slightly (Fig. 21). The amount of the rise was small, but the change in direction was consistent: it increased for seven of nine females in the approaches by them and nine of nine for the approaches to them. In some sense, then, one might say that the time of birth was a time of drawing together of individuals who were relaxed with each other, and that this is one of several dynamics that caused the increase in neighbors described previously. For later months of infant life the rate of noninteractive approaches varied little from month to month, and it was inconsistent in direction of change from individual to individual and month to month. Unlike the rate of interactive approaches I shall consider next, the rate of noninteractive approaches does not reflect developmental changes in mother-infant contact or other aspects of infant maturation.

In examining differences among mothers it is necessary to consider that rates of noninteractive approaches both to and by mothers were probably underestimated, and equally so, in the first few months of this study. Thus to compare the rates for various females I did not include data before late September. Considering the remaining data, I found no pattern in the rate at which various mothers made noninteractive approaches to others. In the simple approaches made to mothers, the three highest-ranking (Vee, Spot, Mom with Moshi) had the highest rates (but Mom with Misty had the lowest). Among the juveniles only Dotty consistently made noninteractive approaches to several different mothers. The lack of a very clear-cut pattern may be due to the overall low rates of these behaviors.

Except perhaps during the first month of infant life, the rate of these calm approaches probably contributes to, but is not sufficient to account for, the appreciable increases that we found in the number of

		by mother	to mother	by mother	to mother	by mother	to mother	by mother	to mother	by mother	to mother	by mother	to mother	by mother	to mother	by mother	to mother	by mother	to mother	by mother	to mother	by mother	to mother	by mother	to mother	by mother	to mother
Adult males	Slim							.23	.39			.40	1.83	.95	1.90									.83	1.09	.31	.92
	Even															.98	1.63										
	Stubby																										
	Ben	.35	.21										.24				.33						.29				
	HT				.22																						
	Max	.21														.65						.35					
	Peter							.39	.23				.48	.38				.44	.44								
	BJ																										
	Chip																										
	Dutch																										
	Red																.65										
	Stiff																										
	Russ																										
	Stu																										
j♂♂ ≥5yrs	Swat																										
	Toto														.57												
	Nog																								.21		
Adult females	Alto																										
	Spot								.54					.38													
	Mom														.38												.31
	Lulu															.33											
	Vee				.28									.95	.38												
	Preg																										
	Scar											.32														.46	
	Oval								.23																	.62	1.23
	Fem																					.29					
	Gin																						.59				
	Jane																										
	Slinky			.22																							
	Handle								.23								.33										
	Plum																									.31	
	Brush							.23																			
	Este																										
	Judy																					.24		.41	.47		
Juveniles <5yrs	Dogo											.32									.28						.62
	Jake																										
	Janet								.39																.26		
	Dotty		.21						.31			.24	.63	.38	.57										.26		.31
	Striper																										.46
	Nazu																										
	Cete												.24														
	Fanny				.22																						
Infants	Pooh																										
	Ozzie																								.57		
	Alice																										
	Fred																										
	Eno																										
	Summer																										
	Pedro																										
	Misty																										
	Bristle																										
	Hans																										
	Grendel																										
	Sesame																										
	Juma																										
	Safi																										
	Moshi																										
	Oreo																										
	Vicki																										
	Peach																										
Total # of acts:		10	17	26	36	11	18	19	49	1	3	27	65	24	25	17	27	3	14	3	9	27	48	47	96	20	35
Total rate per 100 min:		.70	1.19	1.45	2.01	.56	.91	1.48	3.81	.08	.23	2.14	5.16	4.57	4.76	2.91	4.62	.44	2.06	.32	.95	1.59	2.83	2.43	4.96	3.08	5.38

Fig. 20. *Matrix of the rate of noninteractive ("simple") approaches to and by each mother during her infant's first three months of life for which I obtained samples (see text for details). Rates (acts per 100 minutes) are included only for those individuals for whom the values were at least 0.20. However, the last row indicates the total number of acts and the total rate for each mother summed over all individuals.*

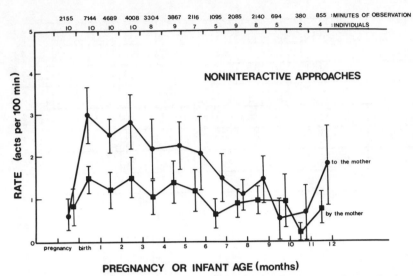

Fig. 21. *Rate of noninteractive approaches to and by mothers during the last month of pregnancy and at each infant age. For each month the rate for each sampled individual was calculated and then the mean and standard error of that mean of the individual values were plotted. See also Chapter 3.*

neighbors; but it is unlikely that the rate itself has an appreciable influence on mothers and infants. What is probably more important is the actor-recipient selectivity of these acts: mothers are brought closer to particular adult males and in some cases to other females with like-aged infants (Fig. 20). Those cases for which we have both likely paternity and behavioral data during the infant's first months indicate that the adult males involved are likely to be the infant's father. The adult female associates are not likely to be closely related through maternal lineages (S. Altmann and J. Altmann 1979), but rather share common needs and experiences, and their infants provide each other with playmates. Actually, mothers of like-aged infants, or the infants themselves, may be more closely related than we suppose from a consideration of only maternal genealogies. Females who are closely associated tend to associate with the same one or two males. Thus their infants are likely to be paternal siblings (J. Altmann 1979). If, in addition, these mothers are of the same age cohort and formed their association in infancy, they, too, may be paternal siblings (see Hrdy 1977). At this stage of our knowledge, however, these suggestions are purely speculative.

INTERACTIVE APPROACHES

A much more common class of approaches consists of approaches that were accompanied by other social behaviors or that were followed by some detectable reaction on the part of the animal being approached (Fig. 22). These interactive approaches were followed by sequences of behavior ranging from relatively long energetic ones, such as fights or other dominance interactions, handling and pulling of associated infants, and grooming, to brief lipsmacking, body aversion, or a slight touch. Most social interactions, both beneficial and harmful ones, begin with an approach. Spatial displacements, which involve approach to another animal's location rather than to the animal itself, are considered separately in the agonistic section below.

In general, mothers initiated interactive approaches slightly more often than they did the simple, or noninteractive, approaches, and did so at a rate that increased very slightly with the birth of their infants and then changed little with the maturation of the infants. This is in striking contrast to the rate of interactive approaches directed toward mothers, which soared with the birth of their infants and then steadily decreased in the subsequent months (Fig. 22). In their infants' first few months of life, mothers were all the recipients of interactive approaches much more often than they were the actors; later the differences were small and reversed in direction for some females and for some months. The six- to eightfold increase in approaches after the birth of an infant is much greater than would be expected from the fact that two individuals (mother plus infant) are now being approached rather than just one, as before.

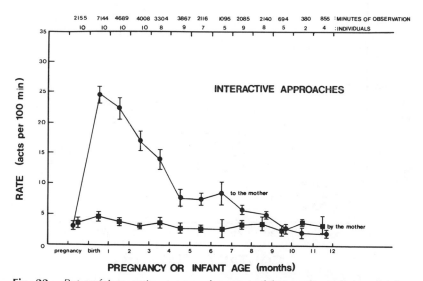

Fig. 22. *Rate of interactive approaches to and by mothers during the last month of pregnancy and at each infant age. Means and standard errors of the mean were calculated as indicated in Fig. 21.*

As an infant matures, its mother may be approached less often because the infant spends less time in contact with her, enabling others to approach the infant at times other than those when it is with its mother. This relationship is quite clear in Fig. 23, in which the rate of interactive approaches is plotted against the time each mother's infant spent in contact that month. The relationship between infant contact and approach rate can be examined at the level of the immediate behavior by determining whether at any given infant age a mother receives more interactive approaches when her infant is actually in contact with her than the expected number of interactive approaches, calculated from the proportion of time the infant spends in contact that month multiplied by the total number of interactive approaches to the mother that month. Using CRESCAT to perform sequential pattern searches, I determined whether the infant was in contact or not when an interactive approach occurred by identifying each change in mother-infant spatial relationships and then labeling each occurrence of a social behavior, including interactive approaches, according to whether the mother and infant were in contact at the time the act was performed. These results are summarized in Table 7.

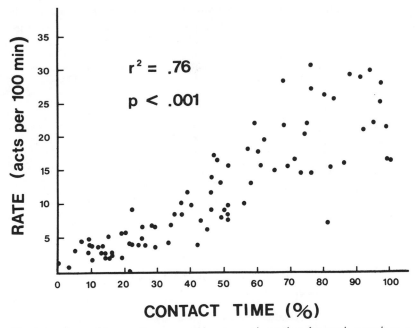

Fig. 23. *Rate of interactive approaches to each mother for each month as a function of the percentage of time that her infant spent in contact that month.*

Table 7. Interactive approaches to the mother when the infant was in contact.

Month of infant life	Number of ♀'s	Mean contact (P)	Mean sample size (number of approaches)	Mean approaches when infant in contact	Number of ♀'s for whom observed = expected	Observed less than expected		Observed more than expected		All data for month pooled	
						Total number ♀'s	Number statistically significant (<.05)	Total number ♀'s	Number statistically significant (<.05)	Observed greater than or less than expected	Significance level
1	10	.95	167.5	158.4	1[a]	4	1	5	2	Less,	$< .001$
2	10	.74	103.3	78.9	1	4	1	5	3	Greater,	$< .001$
3	10	.69	70.1	53.7	1	2	0	7	5	Greater,	$< .001$
4	8	.53	60.3	40.1	1	1	0	6	3	Greater,	$< .001$
5	9	.41	35.9	24.0	0	0	—	9	6	Greater,	$< .001$
6	7	.32	20.7	10.3	2	0	—	5	4	Greater,	$< .001$
7	5	.40	15.4	10.8	0	0	—	5	3	Greater,	$< .001$
8	9	.22	14.1	6.8	1	0	—	8	5	Greater,	$< .001$
9	8	.17	10.9	4.5	3	1	—	5	2	Greater,	$< .001$
10	5	.17	3.4	1.4	1[b]	0	—	3	0	Greater,	$< .02$
11	2	.05	4.5	2.5	1	1	0	1	1	Greater,	$< .001$
12	4	.11	4.7	.25	3		0	0	—	Less,	$> .10$
Total					13	13	2	59	34		

a. One infant was in contact 100% during this month.
b. One female was never approached during this month.

When their infants were in contact, mothers received more inter-active approaches than expected in 59 of 72 mother-months (plus 13 ties). Results of (binomial) significance tests are a function of probabil-ity (contact time) and N (total number of approaches) for any month, which vary considerably in this case. Although the magnitude of the difference between observed and expected values was sufficient to reach the .05 level of significance for only about half (34) of these posi-tive-deviation months (and two of those with negative deviation), the probability of 59 of 72 results being in the same direction (Table 7, last row) by chance is extremely low ($P < .001$). Moreover, if data from all females for each month are pooled, mothers were approached signifi-cantly more often when their infants were in contact during all but months one and twelve (Table 7, last column).

Interest in the mother-infant pair may also wane if older infants are less attractive to others than are younger ones. To examine this possibility I looked at the rate of approaches to a mother when her in-fant was in contact. For four females data were available for the last month of pregnancy as well as for all of the first six months of infant life (but not necessarily for later months). For each mother I calculated the rate at which she received interactive approaches during the last month of pregnancy, which I considered her "base rate," the rate at which she would receive such approaches if she did not have an in-fant. Then, if a female's "attractiveness" was independent of her status as a mother and the presence of her infant, we would expect that she would receive that same base rate of interactions while her infant was in contact (or, at most, twice the base rate if one considers that two individuals are being approached) and that this rate would be inde-pendent of the age of her infant. The results are plotted in Fig. 24, after the base rate for each mother was subtracted from her rate for each month thereafter. They demonstrate that a female received approxi-mately six times as many interactive approaches after parturition when her infant was in contact, and that the rate of approaches was indepen-dent of infant age until the infant was four or five months old. However, the rate dropped dramatically in the sixth month. It is inter-esting to note that at about month six infants have lost most of their distinctive black natal coat. The present results are consistent with the suggestion that infants who have lost their neonatal coat may be less attractive to other group members (see, e.g., DeVore 1963). However, the decrease in approaches to mothers of older infants may be pro-duced by other factors. For example, infants may remain attractive but others may selectively approach them when the infants are not in con-tact with their mothers. In Chapter 9 I shall also examine the contin-

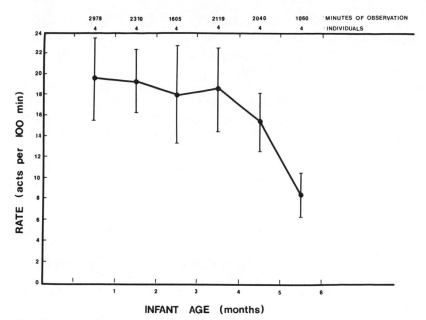

Fig. 24. *Rate of interactive approaches to a mother when her infant was in contact with her. The value used for each mother was the observed value minus her "base rate," the rate at which she received interactive approaches during her last month of pregnancy. The mean base rate was 4.09 with a standard error of the mean of 1.17. Means and standard errors of means were calculated as indicated in Fig. 21. The only individuals used for these calculations were those four for whom data were available for the last month of pregnancy as well as for all of the first six months of infant life.*

gencies between infant contact and maternal activities at certain infant ages, and consider the relationship of those contingencies to alternative explanations for the remaining age effects found in approach rates.

Who were the individuals who approached and interacted with mothers, or whom the mothers themselves approached? For interactions with adult females the pattern was somewhat different from that for interactions with males. A few females seemed to exhibit interest in many infants, others in none. This individual variability is similar to that reported by Breuggeman (1973) for the Cayo Santiago rhesus colony. Thus Spot, Mom, Gin, Slinky, and Oval each appeared as frequent interactants for one-fourth to one-third of the new mothers, themselves of course excluded. At the other extreme, Plum, Scar, Brush, and Lulu never did. The "high-interest" females could also be identified by the amount of time they spent as neighbors of other mothers. None of the high-interest females was closely related to mothers through maternal lineages, and there is no basis for suspecting that through paternal lineages they are more closely related to mothers than are the low-interest females.

New mothers themselves rarely approached others during the early weeks of infant life. One exception was the approach made by mothers to other females with very young infants, an event that was more likely to occur if these other females were lower in rank than the mother (Table 8).

For each mother there were only one or two males that approached and interacted (Fig. 25); often the same males who spent more time near her and with whom she shared noninteractive approaches.

Usually I could predict the existence and identity of such affiliations with males before the birth of an infant, commonly from its mother's mating, grooming, and neighbor associations (Appendix 2). An affiliated male was likely to be the infant's father, that is, the only male, or one of only two or three males, that copulated with the mother during the days that she conceived the infant (Appendix 2).

The older juveniles and subadults, all male, rarely approached the mothers. Among the younger juveniles there was considerable variability. The two highest-ranking juveniles were the ones who most frequently approached and interacted with mothers. However, there was no consistent relationship between juvenile dominance rank and tendency to approach. High-ranking Striper, who appeared as a frequent neighbor of new mothers, was also the juvenile who frequently approached and interacted with mothers, including her own. High-ranking Dotty appeared more frequently as an interactant than she did as a neighbor. Low-ranking Janet and Fanny were intermediate as interactants, and the others rarely approached and interacted.

Also striking from an examination of Fig. 25 is the variability among mother-infant dyads in the number of these frequent interactants. Vee had many (13) such interactants, but most of them were juveniles and older infants. For all the other mothers the variability in the number of interactants was primarily due to the number of adult female interactants, with Gin, Slinky, and Judy all having 6 or more and Spot, Mom, and Vee having fewer than 2 adult females who frequently approached and interacted with them. The highest-ranking mothers had the fewest such interactants: there was a slight tendency for the number of interactants to be inversely correlated with a mother's rank even though there was no tendency for the total rate of interactive approaches to mothers to be rank related. Neither variable was obviously

Table 8. Interactive approaches per 100 minutes for female pairs with young infants.[a]

| Female pairs | | N | X | Y | Magnitude of X relative to that of Y: greater (+) or less (−) | Probability |
Higher-ranking ♀ (month of infant life)	Lower-ranking ♀ (month of infant life)	Total number of acts	Rate of higher- to lower-ranking ♀	Rate of lower- to higher-ranking ♀		
Vee (1)	Plum (1)	7	0.32	0.24	+	.273
Spot (1)	Judy (1)	77	4.58	1.02	+	<.001
Mom-Misty (1)	Brush (1)	7	0.40	0.16	+	.164
Gin (1)	Slinky (1)	52	1.14	2.15	−	.009
Spot (2)	Judy (2)	44	3.69	1.55	+	.003
Mom-Misty (2)	Brush (2)	8	0.62	0.21	+	.109
Gin (2)	Slinky (2)	36	1.98	0.99	+	.018
Preg (2)	Scar (2)	8	1.27	0.42	+	.109
Spot (1)	Slinky (2)	20	0.89	0.48	+	.074
Spot (1)	Gin (2)	12	0.61	0.30	+	.121
Slinky (2)	Judy (1)	24	1.42	0.47	+	.008
Gin (2)	Judy (1)	17	1.15	0.35	+	.018

a. Noninteractive approaches were too rare to analyze in this way, but the limited data exhibited the same trend shown here.

	Mother:	Brush		Gin		Handle		Judy		Mom-Misty		Mom-Moshi		Oval-Oreo		Plum-Peach		Preg		Scar		Slinky		Spot		Vee	
	Months:	(1-3)		(1-3)		(1-3)		(1-3)		(1-3)		(1-2)		(3)		(1)		(1-3)		(2-3)		(1-3)		(1-3)		(1)	
		by mother	to mother	by mother	to mother	by mother	to mother	by mother	to mother	by mother	to mother	by mother	to mother	by mother	to mother	by mother	to mother	by mother	to mother	by mother	to mother	by mother	to mother	by mother	to mother	by mother	to mother
Adult males	Slim								1.48		1.08		2.14												1.14		1.54
	Even																1.63										
	Stubby																										
	Ben		1.82																				1.06				
	HT																										
	Max																										
	Peter										1.16		2.54						2.21								
	BJ																										
	Chip																										
	Dutch																										
	Red		1.12		1.23		1.01																				1.08
	Stiff																										
	Russ																										
	Stu																										
j♂♂ ≥5yrs	Swat																										
	Toto																										1.54
	Nog																										
Adult females	Alto																										
	Spot				1.00				3.42		1.47		2.06		1.14				3.98		2.42		1.18				1.23
	Mom				1.73		1.18		1.17						1.33				1.33		2.56		1.24				1.38
	Lulu																										
	Vee		1.26		1.23				1.48					1.33	1.14								1.53				
	Preg				1.00				1.40	1.24																	
	Scar									1.31								1.03									
	Oval				1.06				1.01										1.62		1.00		2.12			1.08	1.54
	Fem		1.12		1.00												2.28						1.12				
	Gin		1.26				1.72						1.03				1.95				1.42	1.42	1.89				
	Jane																										
	Slinky		1.05	1.28	1.51		1.42		1.17		1.08		1.27				1.14										
	Handle		1.33		1.17												2.11						1.83				
	Plum																										
	Brush					1.32																					
	Este																										
	Judy																		1.03		1.00			3.10			
Juveniles <5yrs	Dogo																										
	Jake																										
	Janet		1.05																		1.14						1.38
	Dotty								1.25				1.03		1.71												1.08
	Striper				1.39						1.47		1.94										1.00		1.24		2.31
	Nazu								1.32																1.03		
	Cete																										1.54
	Fanny				1.23																		1.30				2.15
Infants	Pooh																										
	Ozzie																										
	Alice								1.87																1.03		1.54
	Fred																										
	Eno																										
	Summer																				1.00						1.85
	Pedro																										
	Misty																										
	Bristle																										
	Hans																										
	Grendel																										
	Sesame																										
	Juma																										
	Safi																										
	Moshi																										
	Oreo																										
	Vicki																										
	Peach																										
Total # of acts:		46	327	89	386	100	323	35	348	39	227	36	311	17	80	15	99	24	141	12	168	102	441	126	381	21	184
Total rate per 100 min		3.21	22.85	4.97	21.54	5.06	16.35	2.72	27.08	3.01	17.52	2.86	24.68	3.24	15.24	2.56	16.92	3.53	20.76	1.27	17.83	6.02	26.02	6.51	19.69	3.23	28.31

Fig. 25. *Matrix of the rate of interactive approaches to and by each mother during those of her infant's first three months of life for which I was able to obtain samples. Rates (acts per 100 minutes) are included only for those individuals for which the values were at least 1.00. However, the last row indicates the total number of acts and the total*

related to infant gender or maternal age. Whatever the origin of the differences, some infants had early exposure to interactions with many group members, especially adult females; others had few. Moreover, the infants differed in the particular individuals to which they had the most early exposure, a selectivity that is not easily related to kinship, as discussed previously.

The nature of the social interactions in which mothers and infants are involved may provide important clues to differences in early experience. I shall now examine several major types of social interactions, some apparently beneficial, others stressful, in more detail. These interactions sometimes followed immediately upon the approaches I have just discussed. At other times they occurred between two animals that already were in proximity to each other.

Grooming

Assumed to be beneficial (by, e.g., Hutchins and Barash 1976), grooming is the most obvious and time-consuming form of primate social behavior. Table 9 and Fig. 26 indicate the amount of time that grooming was done to or received from others for the females in this study. The values were obtained by using CRESCAT to locate grooming bouts in the files and then having the computer subtract the time that grooming ended from the time that it started, cumulate these times for each dyad during a month, and divide the cumulated time by the appropriate total observation time for the month. Females spent little time either being groomed (1 to 2 minutes per 100) or grooming (3 or 4 minutes per 100) during the last month of pregnancy and were more often actors than recipients of grooming at that time (six of the females groomed more than they were groomed, three tied, and one, Plum, received more grooming than she performed). During the first month of infant life the average rate of grooming by mothers decreased to between 1 and 2 minutes per 100 and the amount of grooming time that mothers received increased greatly (to 7 minutes per 100), further increasing to a peak of over 8 minutes per 100 during the second month of infant life. The amount of grooming mothers received dropped greatly during the third month, to 3 to 5 minutes per 100, and decreased very slightly each month thereafter to levels of about 2 to 3 minutes per 100 (sample sizes became small and variability great). After their infants were two months old mothers tended to spend about 2 minutes per 100 grooming others.

All mothers except Spot groomed their own infants more than they groomed any other individual (Table 9), and all were groomed by others more than they groomed others (Table 9, Fig. 26). As new

Table 9. Grooming of and by mothers during months one through three.
Mothers are ordered by dominance rank at the birth of the infant.
Individuals are listed by time spent per 100 minutes and percentage
for those who received or provided at least 10 percent of a
mother's grooming. "All" indicates total time for all group members.

Mother-infant	Other individual	Grooming of mothers		Grooming by mothers	
		Time spent	Percentage	Time spent	Percentage
Spot-Safi	Janet	1.65	32%	—	—
	Slinky	1.28	25%	—	—
	Fanny	.53	10%	—	—
	Slim	—	—	1.42	23%
	Judy	—	—	.82	13%
	High Tail	—	—	.76	12%
	Safi	—	—	1.49	24%
	All	5.15		6.10	
Mom-Moshi	Janet	1.70	16%	—	—
	Nazu	1.67	16%	—	—
	Striper	1.25	12%	.63	23%
	Slinky	1.24	12%	—	—
	Peter	—	—	.50	18%
	Slim	—	—	.36	13%
	Moshi	—	—	1.30	47%
	All	10.46		2.79	
Mom-Misty	Peter	1.48	47%	1.09	46%
	Slinky	.70	22%	—	—
	Janet	.49	15%	—	—
	Slim	—	—	.83	35%
	Misty	—	—	.37	16%
	All	3.15		2.39	
Vee-Vicki	Lulu	1.58	17%	—	—
	Preg	1.48	16%	—	—
	Cete	1.41	15%	—	—
	Nazu	1.20	13%	—	—
	Striper	1.14	12%	—	—
	Slim	—	—	1.79	39%
	Vicki	—	—	2.55	56%
	All	9.16		4.55	

Table 9 (*continued*).

Mother-infant	Other individual	Grooming of mothers		Grooming by mothers	
		Time spent	Percentage	Time spent	Percentage
Preg-Pedro	Nazu	.57	44%	.14	13%
	Lulu	.43	33%	—	—
	Stubby	.30	23%	.82	73%
	Judy	—	—	.15	14%
	Pedro	—	—	.01	1%
	All	1.30		1.13	
Scar-Summer	Stubby	1.47	33%	.75	49%
	Vee	.67	15%	—	—
	Slinky	.60	14%	—	—
	Judy	.53	12%		
	Slim	—	—	.45	30%
	Summer	—	—	.24	16%
	All	4.39		1.52	
Oval-Oreo	Janet	2.62	74%	2.19	60%
	Lulu	.70	20%	.68	19%
	Oreo	—	—	.54	15%
	All	3.53		3.64	
Gin-Grendel	Janet	1.42	18%	—	—
	Lulu	.90	12%	—	—
	Handle	.80	10%	—	—
	Red	—	—	.66	15%
	Grendel	—	—	2.05	48%
	All	7.81		4.31	
Slinky-Sesame	Plum	1.36	17%	—	—
	Vee	1.30	16%	—	—
	Fanny	1.14	14%	—	—
	Janet	.79	10%	—	—
	Spot	—	—	.38	19%
	Sesame	—	—	1.08	54%
	All	8.18		2.00	

(*continued*)

Table 9 (*continued*).

Mother-infant	Other individual	Grooming of mothers		Grooming by mothers	
		Time spent	Percentage	Time spent	Percentage
Handle-Hans	Gin	3.45	39%	.41	11%
	Slinky	.87	10%	.48	13%
	Scar	—	—	.38	10%
	Hans	—	—	1.16	32%
	All	8.81		3.66	
Plum-Peach	Este	2.11	37%	—	—
	Gin	1.17	21%	—	—
	Max	1.06	19%	—	—
	Preg	.59	10%	—	—
	Even	—	—	.52	10%
	Peach	—	—	4.43	88%
	All	5.67		5.06	
Brush-Bristle	Judy	1.03	34%	—	—
	Max	.47	16%	.07	13%
	Slinky	.45	15%	—	—
	Este	—	—	.08	15%
	Even	—	—	.06	11%
	Plum	—	—	.06	10%
	Bristle	—	—	.21	38%
	All	2.97		.56	
Judy-Juma	Janet	2.70	47%	.77	34%
	Oval	1.37	24%	.26	12%
	Peter	1.18	20%	.22	10%
	Juma	—	—	1.00	44%
	All	5.79		2.27	

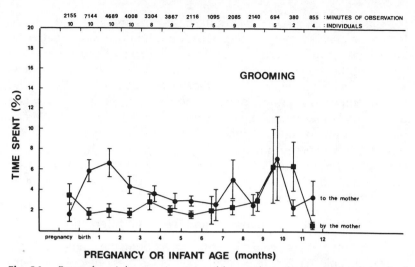

Fig. 26. *Rate of social grooming to and by mothers during the last month of pregnancy and at each infant age. Means and standard errors of means were calculated as indicated in Fig. 21.*

mothers, only Brush and Handle did much grooming of other adult females. For other new mothers almost all their grooming was directed to their own neonates, their juvenile offspring, and one or two associated males. These males and juvenile offspring also groomed the mothers, as did a few other associates (Table 9). The identity of these other groomers differed among mothers with the striking exceptions of adult female Slinky and juvenile female Janet, who did appreciable grooming of most of the new mothers. For Janet, this included providing most of Oval's grooming, which Oval reciprocated. Strangely, Oval's own juvenile daughter, Fanny, neither groomed nor was groomed much by her mother when Oreo was young, although they were frequent grooming partners at other times.

There were no consistent relationships between a new mother's dominance rank and the amount of grooming done by or to her (Table 9). This contrasts with the finding of Seyfarth (1977) for a group that included eight adult female chacma baboons. The other baboon studies have not mentioned a correlation between maternal rank and the amount of grooming received. Except for that of Saayman (1971), they do report more grooming of females with young infants than of females in most other reproductive stages.

Grooming is usually assumed to be beneficial for removal of ectoparasites and for social bonding—itself of unknown biological cost or benefit. Although grooming is quite time-consuming, it is done primarily to, not by, mothers (Fig. 26) and, moreover, is initiated by others: that is, mothers rarely approach and solicit (i.e., present for) grooming. Groomees are in general quite relaxed, although mothers are somewhat less relaxed than other groomees because those who groom mothers often stop grooming and reach for the infant, at which point a mother responds immediately by clutching her infant. We can only say at this point that grooming, unlike some other interactions, probably is not harmful and may be beneficial. Although it is time consuming, and from that standpoint would seem to stress mothers' time budgets additionally, time spent being groomed may obviate the need for some of the usual rest time.

Dominance Relationships and Agonistic Interactions

The outcomes of agonistic interactions during the whole study are shown in Table 10. For almost every pair of females one or both members of the pair experienced at least two different reproductive conditions during the year. However, no pair of individuals reversed the direction of their dominance-subordinance relationship as a result of birth or maturation of their infants (Table 10), that is, dominance relationships were independent of reproductive states (Hausfater 1975a and Hausfater et al. in preparation, Rowell 1966a, Seyfarth 1976). For possible changes in agonistic relationships due to the birth of an infant we must look beyond the simple "winner" or "loser" outcome of agonistic encounters. Some other aspects of these interactions are considered below.

SPATIAL DISPLACEMENTS

Sometimes one individual moves closer to another individual (or would have done so if the second animal did not leave), but it is the space or location of the second animal that is being approached, rather than the animal occupying that space. These spatial displacements, or supplantations, are usually distinguishable from what I have called approaches by the direction of gaze of the approaching animal and the endpoint of the approach: spatial displacements do not involve "tracking" or following the recipient. The animal whose site is approached responds with mild submissive gestures (Appendix 4), for instance, a slight cower or several rapid glances, suddenly discontinues its current activity, and moves away. These spatial displacements were not considered approaches to an animal in the previous discussions. Supplantations were commonly but not necessarily followed by the actor's assuming the same activity (feeding, grooming) that had occupied the

Table 10. Adult female dominance matrix, based on all agonistic encounters observed during ad libitum samples and on displacements observed during focal samples. Data for the entire study, regardless of female reproductive status, are included.

Winner	Loser																
	Alto	Spot	Mom	Lulu	Vee	Preg	Scar	Oval	Fem	Gin	Jane	Slinky	Handle	Plum	Brush	Este	Judy
Alto[a]	—	5	3	2	1	4	10	3	0	2	1	3	3	3	11	4	1
Spot		—	1	14	14	35	14	15	1	7	0	13	16	8	18	9	8
Mom		1	—	9	10	33	32	14	12	19	0	15	14	10	24	10	7
Lulu				—	2	4	3	6	2	5	1	10	19	4	10	4	0
Vee					—	5	15	13	3	15	0	12	22	4	17	2	0
Preg						—	5	8	6	9	1	7	13	3	13	0	3
Scar			1				—	10	2	13	5	6	8	4	18	4	2
Oval							1	—	2	11	0	10	18	5	25	3	1
Fem								1	—	12	0	22	24	11	15	2	2
Gin						1		2	1	—	0	33	42	17	16	6	2
Jane[b]											—	0	0	0	2	2	0
Slinky										1		—	25	11	18	4	8
Handle													—	18	26	13	9
Plum														—	22	7	5
Brush															—	4	8
Este																—	6
Judy[c]																	—

a. Alto died 21 May 1976.
b. Jane died 23 Oct. 1975.
c. Judy died 10 May 1976.

animal who was displaced and doing so at the same spot. Actors thereby take over resources such as partially dug grass corms (S. Altmann 1974, Post 1978). "Approach-avoidance" interactions are frequently discussed in the primate literature, as are spatial displacements, but definitions are rarely provided and there has not been a tradition of consistent terminology. I have used a somewhat more restrictive definition of spatial displacements than Post did; he would include as displacements some of the approach-avoidance interactions that I discussed previously in the sections on approaches. Rowell (1966a, 1969) and Seyfarth (1976) made a different pooling of categories, combining and considering as "friendly" approaches (Rowell 1966a, 1969) all approaches to the place or to the individual that did not include any threats by the approacher.

Spatial displacements are a common type of agonistic encounter (Hausfater 1975a). Averaged over all females, the rate of spatial displacements both to and by mothers was slightly lower during the first month of infant life than during the last month of pregnancy, but otherwise the average remained at fairly constant levels independent of the existence or age of a female's infant (Fig. 27). Six of nine females were displaced at lower rates during their infants' first month of life than during the last month of pregnancy (Fig. 28, top). The three whose rates

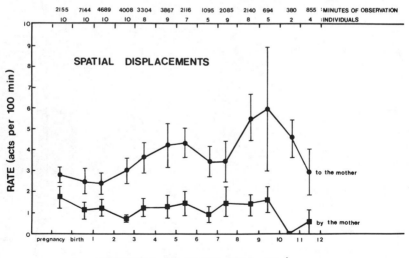

Fig. 27. *Rate of spatial displacements (supplantations) to and by mothers during the last month of pregnancy and at each infant age (see text for details). Means and standard errors of means were calculated as indicated in Fig. 21.*

increased were three of the four females whose infants were born during the latter part of the dry season. Post (1978) found that displacement rates in general are higher during that season. Unlike berries, flowers, and most other baboon foods, the main dry-season food, grass corms, requires considerable time and energy to process, because the corms are dug from the ground. Thus one animal can benefit appreciably by supplanting another from a corm that the second has been digging. From month one to two of infant life the rate of spatial displacement increased for six of nine females, as it did for six of nine between months two and three of infant life. By month three, five of eight females were supplanted at higher rates than they had been during the last month of pregnancy. Thus infant birth may bring with it slight immunity against spatial displacements, but this immunity has disappeared by the infant's third month.

Again, Brush and Mom provide an interesting contrast. During the month before Moshi was born, high-ranking Mom's rate of spatial displacements was approximately the same as Brush's was during her last month of pregnancy. However, Brush was displaced at a slightly higher rate during her infant's first month of life, whereas Mom was displaced appreciably less after Moshi was born. Even if we take into account the fact that no observations are available for Moshi's third month, Brush was displaced about twice as often as was Mom during the first few postpartum months.

During the last month of pregnancy high-ranking females as a group were supplanted at much lower rates than were the low-ranking ones, but there was no consistent tendency for rank order and rate of spatial displacements to be related (Fig. 28, top). The decrease in rate of displacements during the next month resulted in less difference in rate of displacement between high- and low-ranking females during their infants' first month than there had been before their infants' birth. This perhaps accounts for DeVore's conclusion that new mothers gained in dominance rank (Hall and DeVore 1965). Although DeVore's matrix of dominance interactions did not suggest a rank change (change in outcome of agonistic interactions) as a function of reproductive state (Hall and DeVore 1965), DeVore indicated that rank changed when a female's infant was born. Perhaps this interpretation was the result of his perceiving reduced rates of spatial displacements at this time. However, as noted above, in the present study even this possible immunity was brief and had entirely disappeared two months later. Also, in the present study low-ranking mothers were displaced more often than high-ranking ones. That is, there was a general trend for maternal rank to be inversely related to displacement rate pooled

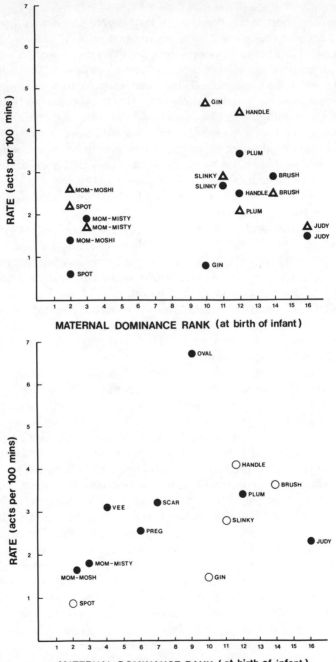

over the first three months of infant life (Fig. 28, bottom), the time when overall interaction rates were by far the highest. Somewhat more clear-cut are the data for the five females for whom several sample days are available for each of these three months, as indicated on the same graph (Fig. 28, bottom). Using γ, a measure of association (Goodman and Kruskal 1954) indicating how much more probable it is that a pair of individuals will be ranked in the same, rather than the opposite, order on two variables (in this case dominance rank and rate of being supplanted), I obtained values of .40 for all females but .80 for just those five females. That is, pairs of mothers were much more likely than not to have the same relative dominance rank and rank order on rates of supplantation.

OVERT AGGRESSION

It was extremely rare (less than once per 100 minutes) for mothers either to display or to receive acts of aggression or overt threat. Of the 192 cells (mothers × months observed × 2 — actor or receiver), in only 10 was the rate over two acts per 100 minutes; in only 1 was it over four. In general, these females were more likely to be the recipients of aggression than the performers of aggressive acts. This was true for the last month of pregnancy and throughout the first year of infant life, with no consistent variability as a function of infant age. There were no consistent patterns of rates of aggression with respect to dominance rank. The absence of detectable developmental and rank differences is probably related to the general rarity of the behaviors.

DISTRESS

If females receive a considerable amount of grooming and little overt aggression when they have young infants, is it reasonable to view their increased social involvement as basically calm or "positive"? Perhaps not, as suggested by the reports of mothers frequently avoiding "friendly" approaches (e.g., Rowell et al. 1968, Seyfarth 1976). I analyzed the data on behaviors of "distress" or "fear" during social inter-

---

Fig. 28. *Rate at which mothers were supplanted as a function of the mother's dominance rank at parturition. Top: rate during the last month of pregnancy (△) and the first month of infant life (●), only for females for whom data were available for both months. Bottom: average rate for the first three months of infant life; females for whom data were available for any of these months (●), $\gamma = 0.40$ (Goodman and Kruskal 1954), and those for whom at least two days' data were available for each of these three months (○), $\gamma = 0.80$ (Goodman and Kruskal 1954).*

Fig. 29.   *Embracing her new infant Peach, Plum turns away, averting her gaze and giving a mild cower, as Handle tries to investigate the infant.*

actions. These social interactions do not include, but rather are in addition to, the spatial displacements that have already been discussed. I counted one occurrence of a distress response whenever a female gave any one or more of the submissive behaviors listed in Appendix 4, regardless of the intensity or duration of such behavioral acts; that is, if a mother responded to a partner with several submissive behaviors such as a grimace and a tail-up, I counted all the behaviors as one distress response unless an act by the partner intervened. Thus, for the time being, I have ignored intensity of distress.

Figs. 30 through 32 indicate that the rates of distress responses averaged less than 3 per 100 minutes (Fig. 30) during the last month of

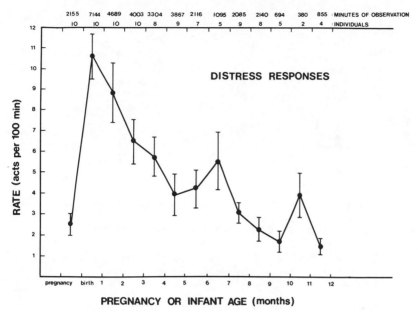

Fig. 30. *Rate of distress responses by mothers during the last month of preg-nancy and at each infant age. Means and standard errors of means were calcu-lated as indicated in Fig. 21.*

pregnancy and were only slightly related to maternal dominance rank (Fig. 31). However, during the first month of infant life, mothers averaged almost 11 distress responses per 100 minutes (Figure 30); in-dividual response rates were strongly related to maternal rank (Fig. 30), with low-ranking mothers having higher rates of distress responses than did higher-ranking ones (Fig. 31, top). The same result, indicating the importance of dominance rank, is seen in the pooled data for months one through three (Fig. 31, bottom), $\gamma = .80$ (Goodman and Kruskal 1954); that is, the birth of an infant greatly exaggerated the ef-fects of dominance rank and dominance affected the amount of distress experienced by new mothers and, in all likelihood, by their infants.

The rate of maternal expressions of distress decreased with in-creasing infant age until by month eight levels were near those of late pregnancy. An alternative way of viewing the effect of the infant on stressful social interactions for its mother is to plot these interactions for each mother for each month against the percentage of time the mother's infant was in contact that month (Fig. 32), as was done for interactive approaches, again producing a striking correlation but with

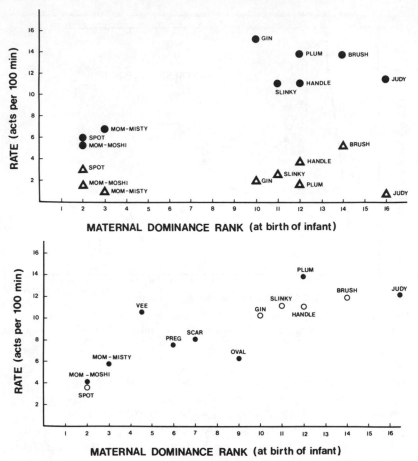

Fig. 31. *Rate of distress responses by mothers as a function of the mother's dominance rank at parturition. Top: rate during the last month of pregnancy (△) and the first month of infant life (●), only for females for whom data were available from both months. Bottom: average rate for the first three months of infant life; females for whom data were available for any of these months (●), γ = 0.80 (Goodman and Kruskal 1954), and those for whom at least two days' data were available for each of these three months (○).*

much more scatter because dominance rank did not affect the rate of approaches as it did the rate of distress responses.

The fact that the rate of distress responses by mothers drops off at about the same rate as does mother-infant contact suggests that mothers may be subjected to stressful situations just when their infants are in contact, or perhaps even that they could avoid stressful situa-

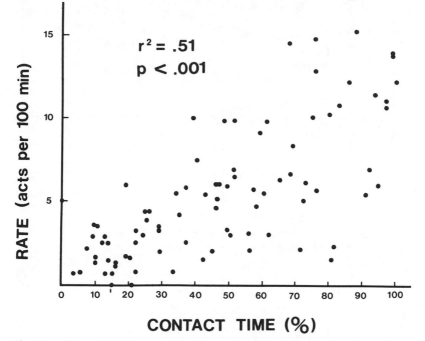

$r^2 = .51$
$p < .001$

Fig. 32. *Rate of distress responses by each mother for each month as a function of the percentage of time that her infant spent in contact that month.*

tions by not being in contact with their infants. To investigate this possibility I analyzed the rate of distress responses when the mother's infant was in contact, as I had done for the interactive approaches to mothers. If infant presence is the relevant situational variable, one would predict that mothers would express distress disproportionately when their infants were in contact. Moreover, if the infant's immediate presence alone was what resulted in the mother being in stressful situations, regardless of infant age, the rate of distress responses when the infant was in contact would be expected to remain constant as the infant matured. Alternatively, if an older infant is less attractive to others and/or if a mother is less sensitive or fearful when her infant is older, then the overall decline in distress would also be present when mother and infant were in contact.

I first analyzed the rate of distress responses, contingent on the infant being in contact at the time of the interaction, using conditional pattern searches in CRESCAT, as described previously. As can be seen in Table 11, mothers did perform behaviors of distress disproportionately when their infants were in contact, particularly when their infants

Table 11. Distress responses of mothers when the infant was in contact.

Month of infant life	Number of ♀'s	Mean contact (P)	Mean sample size (number of distress responses)	Mean distress responses when infant in contact	Number of ♀'s for whom observed = expected	Observed less than expected		Observed greater than expected		All data for month pooled	
						Total number ♀'s	Number statistically significant (<0.05)	Total number ♀'s	Number statistically significant (<.05)	Observed greater than or less than expected	Significance level
1	10	.95	74	72.4	1	2	0	7	2	Greater,	<.001
2	10	.74	40.9	36.4	0	2	0	8	5	Greater,	<.001
3	10	.69	25.2	20.9	0	0	—	10	3	Greater,	<.001
4	8	.53	24.5	17.9	0	1	0	7	3	Greater,	<.001
5	9	.41	18.4	11.1	0	2	0	7	2	Greater,	<.001
6	7	.32	12.6	6.6	0	0	—	7	1	Greater,	<.001
7	5	.40	10.6	4.6	0	2	0	3	0	Greater,	>.1
8	9	.22	7.1	2.9	0	4	0	5	2	Greater,	<.001
9	8	.17	6.8	2.0	0	5	0	2	1	Less,	<.02
10	5	.17	2.2	.4	1	2	0	2	0	Greater,	<.1
11	2	.05	7.0	1	1	0	—	1	0	Greater,	<.1
12	4	.11	3.3	.25	0	3	1	1	0	Less,	>.1
Total					4	23	1	60	19		

were less than six months old. The data partitioned by individual mothers and by months rarely reached statistical significance, but the direction of the results is quite consistent; in 60 of 83 mother-months (plus 4 ties), infant contact was linked to maternal distress (P < .05, binomial test). Pooling the data for all females for each month results in significant values for months one through six and months eight and nine (Table 11, last column).

As I had done with interactive approaches, I then examined the rate of distress responses when a mother's infant was in contact for those mothers for whom data were available for the last month of pregnancy and also the first six months of infant life (Fig. 33). These females averaged approximately four distress responses per 100 minutes during the last month of pregnancy. When their infants were in contact the following month, there was an average increase of more than nine

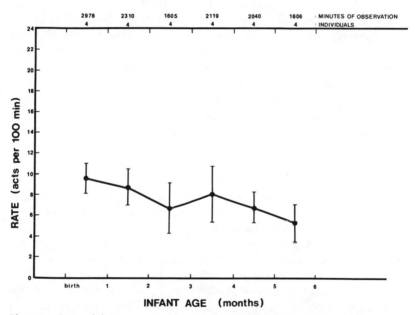

Fig. 33. *Rate of distress responses by a mother when her infant was in contact with her. The value used for each mother was the observed value minus her "base rate," the rate at which she gave distress responses during her last month of pregnancy. The mean base rate was 3.83 with a standard error of the mean of 0.83. Means and standard errors of means were calculated as indicated in Fig. 21. The only individuals used for these calculations were those four for whom data were available for the last month of pregnancy as well as for all of the first six months of infant life.*

such responses per 100 minutes. As expected from the several previous findings, the rate of maternal distress when the infant was in contact decreased only slightly with increased infant age for the first few months of infant life. That is, mothers are, in general, put in stressful situations because they are with their infants, and this decreases relatively little during their infants' first few months of life. Although the absolute rate of distress responses during infant contact changed little during the first six months, the percent decline was more than for interactive approaches to the mother and, with the exception of an increase in the fourth month of infant life, there was a steady decrease in the rate for each of the first six months. This slow steady decrease continued for those two mothers for whom I also had continuous data for the next several months (months six through nine). The gradual reduction in distress responses during contact that occurs may be due either to the mother's decreased "feeling" of vulnerability when the infant is older or to reduced attractiveness of the infant (DeVore 1963), which reduces other animals' attempts to interact. However, the rate of distress responses when the infant was in contact decreased with infant age slightly earlier and more steadily than did the rate of interactive approaches to the mother (a rate that variably increased or decreased very slightly until month five) and showed no sharp change at any infant age; this suggests that the mother's feeling of vulnerability or evaluation of the potential danger in the interactions may decrease earlier and respond to gradual infant maturation, as contrasted with the behavior of others, which changed dramatically after the major loss of the natal coat.

Mothers directed the distress responses and acts of embracing or restraining of their infants disproportionately more toward other adult females than toward individuals of other age-sex classes, and especially toward those females who were both higher in rank than the mother and among the females particularly "interested" in infants, especially Spot, Mom, Gin, and Slinky (Fig. 34). Thus only male Slim and, to a lesser extent, juvenile male Nog caused appreciable distress to Spot, and only Spot and to a lesser extent Slim and male Peter did so to Mom. In contrast, low-ranking Brush often was distressed by male Red and by females Spot, Vee, and Slinky; she was somewhat less distressed by Fem and Gin.

Infant Handling and Pulling

What are some of the distress-provoking situations to which mothers are exposed? Recall that in the analysis of aggressive behavior, approach was not considered a threat, nor have previous re-

Fig. 34. Matrix of the rate of distress responses to and by each mother during her infant's first three months of life for which I obtained samples. Rates (acts per 100 minutes) are included only for those individuals for whom the values were at least 1.00. However, the last row indicates the total number of acts and the total rate for each mother summed over all individuals.

	Total # of Acts	Total Rate per 100 min
	66	10.63
	36	8.28
	16	4.27
	71	7.83
	15	2.78
	7	2.03
	72	9.29
	59	8.94
	40	7.41
	32	5.38
	30	6.90
	3	1.18
	22	3.55
	26	4.81
	0	0.00
	25	3.40
	9	1.71
	15	2.86
	41	6.67
	11	6.11
	17	5.16
	20	6.81
	14	3.42
	44	6.52
	35	5.9
	19	5.51
	27	3.46
	7	1.73
	13	1.73
	44	6.77

searchers considered it such. Nor has handling or pulling of an infant appeared in catalogues of threat or aggression, and thus I did not include them as such in the analysis of aggression above. Yet a mere approach often elicits submissive (including avoidance) behaviors, as several authors have noted (e.g., DeVore 1963, Rowell et al. 1968, Seyfarth 1976). In addition, other animals often handle an infant, sometimes even pulling it away from its mother with sufficient force to produce counter-pulling by the mother, screeching by the infant, and other signs of distress in both (Fig. 35 and see definitions in Appendix 4). Such pulling and handling occurred often during the infant's first month, was in addition to mere touching or muzzling, and was primarily done by other adult females to infants of lower-ranking females. In this study it seemed clear that mothers "perceived" the mere approach or presence of certain individuals, and certainly handling and pulling of the infant, as a threat or a source of distress. In the extreme cases, snatching resulted in infants being kidnapped. Usually this kidnapping was brief, but in one instance the kidnapper, Gin, kept the infant overnight. When female Handle finally was able to regain possession of her two-day-old son about 15 hours after kidnapping, Hans was weak and appeared dehydrated. Hrdy (1976) reviews instances of kidnapping in several species.

Repeated handling and pulling of infants of low-ranking mothers seemed to have at least two consequences. It was probably a major factor affecting maternal behavior, which I shall discuss in Chapter 7. In addition, some mothers, when they were repeatedly approached and their infants repeatedly yanked, would approach and sit next to a particular adult male associate. Thus, as suggested in previous sections, relationships with adult males provided a major source of variability in infant experience. These relationships will be discussed below.

Associated Adult Males

Adult males, like females, showed varied degrees of interest in infants. However, for males more than for females, interest depended strongly on the identity of the mother (Ransom and Ransom 1971; Table 12). Similarly, each mother's tolerance of a particular male's approaches depended on the identity of the male (Table 12). That is, there were specific adult male–mother associations, and these seemed

Fig. 35. *Three frames from a long sequence in which female Mom persisted in pulling various parts of infant Grendel's body while his mother, Gin, held him.*

Table 12. Males associated with mothers during the first three months of infant life. Mothers are ordered alphabetically. Column entries indicate relationships meeting or exceeding criterion levels.[a]

Mother-infant	Males by dominance rank[b]	Neighbor	Noninteractive approaches		Grooming		Total
			♂ approached mother	Mother approached ♂	♂ groomed mother	Mother groomed ♂	
Brush-Bristle	Ben	X	X	X			3
	Even					X	1
	Max	X		X	X	X	4
	Even	X					1
Gin-Grendel	High Tail		X				1
	Red	X				X	2
Handle-Hans	Even	X					1
Judy-Juma	Slim	X	X	X			3
	Peter	X	X	X	X	X	5
Mom-Misty	Slim	X				X	2
	Peter	X			X	X	3
Mom-Moshi	Slim	X	X	X		X	4
	Ben		X				1
	Peter	X	X			X	3

Mother–Infant	Male	Neighbor	Grooming	Noninteractive approaches	Days of rank change
Oval-Oreo	Slim	X	X		3
	Peter	X			1
Plum-Peach	Even	X	X	X	4
	Ben	X	X		1
	Max		X	X	2
Preg-Pedro	Red	X	X		2
	Slim	X	X		2
	Stubby	X	X	X	2
	Peter	X	X		3
Scar-Summer	Slim	X		X	1
	Stubby	X	X	X	3
	Peter	X			1
Slinky-Sesame	Even	X			1
	Ben	X	X		2
	Max	X			2
Spot-Safi	Slim	X	X	X	4
	High Tail		X	X	1
Vee-Vicki	Slim	X	X	X	4

a. Neighbor: within 2 meters at least 5 percent of the time; noninteractive approaches: rate of at least 0.20 times per 100 minutes; grooming: at least 10 percent of mother's grooming. See also Table 9.

b. Males ordered by dominance rank on infant's birth date. For days of rank change, see Table 19.

Table 13. Relationships between adult males and mothers or their infants. Mother-infant pairs are ordered by infant birth dates. A blank indicates no relationship, X indicates a relationship, and ? indicates no available information.

Mother-infant	Males by dominance rank	Likely father	Associate of mother before infant's birth	Neighbor during months 1–3[a]	Male presence reduced		Later associate of infant
					Pulling of infant, month 1	Infant contact, month 2	
Plum-Pooh	Chip		?				X
	Max		?				X
Oval-Ozzie	Slim[b]	X	?	?			
	High Tail[b]	X[c]	?	?			X
Alto-Alice	Slim	X	?	?			
	Peter	X	X	?			X
	BJ	X	?	?			
	Crest	X[c]	?	♂ left group			
Fem-Fred	Max	X	?	X[d]	?	?	X
	BJ	X	?	?	?	?	X
Este-Eno	BJ[e]	?	?	?			
Scar-Summer	Stubby	?	?	X(3)	?	?	♂ died
	Peter	?	?	X(1)	?	?	
Preg-Pedro	Slim	?	?	X(2)	?	?	
	Peter	?	?	X(3)	?	?	
Mom-Misty	Slim	?	?	X(2)			Infant died
	Peter	?	?	X(3)			
Brush-Bristle	Ben	?	X	X(3)	X	X	X
	Max	?	X	X(4)		X	
Handle-Hans	Even	?		X(1)	X	X	X
Gin-Grendel	Even	?[f]		X(1)	X	X	
	Red	?[f]	X	X(2)	X	X	♂ left group
	High Tail[g]	X					X

Pair	Male					
Slinky-Sesame	Even	X		X(1)		
	Ben	X		X(2)	X	X
	Max[h]	X		X(2)	X	X
Judy-Juma	Stubby		X	♂ died		–
	Peter	?[f]	X	X(5)	X	X
	Slim	?[f]		X(3)		X
Spot-Safi	Slim	?[f]		X(4)		X
	Stubby	X	X			X
Mom-Moshi	Slim	X		X(4)		X
	Peter			X(3)	X	X
Oval-Oreo	Slim	X		X(3)		
	High Tail	X	X		X	
Vee-Vicki	Slim	X		X(4)	X	Infant died
	Red	X	X			
Plum-Peach	Even	X[c]		X(4)		Infant died
	Max[i]	X		X(2)		–
	Red	X		X(2)		–
	High Tail	X		♂ died		–

a. The criterion value for neighbor relationship was presence of the male within 2 meters at least 5 percent of the time. The number in parentheses indicates the number of relationships observed (Table 12, last column).

b. Slim and High Tail were Oval's consorts after postpartum amenorrhea.

c. The most likely father, as discussed in the text and Appendix 2.

d. Data from ad libitum samples used.

e. BJ and High Tail were Este's consorts after postpartum amenorrhea.

f. Incomplete consort records for period of conception.

g. The infant unsuccessfully tried to establish other relationships after High Tail's death.

h. Max spent much time with Pooh.

i. Max stayed behind at infant birth.

to reflect specific relationships between male and female that were established before the infant's birth (Appendix 2; Table 13). In most cases, at least two males attempted to associate with the mother and infant soon after parturition. The higher-ranking male sometimes exhibited herding behavior, followed the mother, threatened other males away, groomed the mother, and, in fact, exhibited all elements of a sexual consortship except for actual mounting. This was most striking in Slim's pursuit of Mom when Moshi was born. He frequently herded Mom from Peter and occasionally even fought with Peter in these situations. Likewise, Ben prevented Max from remaining close to Brush when Bristle was born. Wounded and dropping in rank, Ben was unable to do the same after Sesame's birth to Slinky in February. As with sexual consortships, the role of the female in these interactions was not that of a passive observer. A mother did not follow a lower-ranking male while being herded by a higher-ranking one, nor did she ever threaten males who followed her; but she selectively avoided certain males when they approached, reciprocated the following or not, selectively followed and groomed certain males, and so on.

The result was that some relationships between mothers and adult males were more reciprocal and enduring than were others, as can be seen in Table 12 and 13. Table 12 is a summary table that was prepared by using the data for grooming and noninteractive approaches from Table 9 and Fig. 20. Rather than use the neighbor data from Fig. 18, I used pooled data over the first three months to make them more comparable to those in the other columns. Interactive approaches were not included because these often involve agonistic as well as affiliative or neutral interactions, as discussed previously.

Table 13 summarizes the available information on the enduring nature of some relationships and on details of the effect of male presence near mothers and their infants. The associated male, either through overt behavior or through his mere presence, often provided a buffer between the mother-infant dyad and other group members. When the male was near the mother-infant pair, others approached more hesitantly; they gave repeated "anxious" glances toward the adult male, and veered in their approach to the mother so that they approached her on the side opposite the male. They sat farther from the pair than usual, watching, and then suddenly trotted toward the mother and infant as soon as the male moved away. The male sometimes overtly threatened those who approached the pair, particularly if the interloper pulled at the infant. One result of such male influence was that the rate of infant pulling during an infant's first month was usually lower in the presence of the associated male (Table 13, column

5). Note that the presence of Slim did not usually reduce the rate of infant pulling. This was because he was the major infant puller for those females with whom he associated. An additional effect of male presence was that in the second month of life, when most infants spent moderate amounts of time away from but still near their mothers, infants were out of contact more when the male was within 2 meters than they were otherwise (Table 13, column 6).

Adult male-infant relationships persisted beyond the neonatal period (Table 13, column 7). Older infants rested against their male associates and obtained rides from them (Fig. 36), ran to them in times of distress, and took greater liberties with these males than with others by feeding near them or by getting meat scraps from them. Elsewhere (J. Altmann 1978) I have referred to the male who had an enduring and beneficial relationship with an infant as its "godfather". Particularly striking godfather relationships were those of Pooh with Chip and then later with Max, Fred with Max, Alice with Peter, Bristle with Ben, Hans with Even and Grendel with High Tail.

There is another side to the male-infant relationships: the males sometimes derived apparent benefits from their relationships with infants. In several species males under attack sometimes take infants to their ventrum, an act thought to serve as an "agonistic buffer" (see Deag and Crook 1971, Ransom and Ransom 1971), that is, to mitigate attack. However, in this study, only if the male and the infant had an existing positive relationship did the infant remain riding and possibly serve this function when the male was under attack. Otherwise the infant usually refused to cling. Then it either dropped off or the male embraced it continuously, running three-legged. The male's locomotion was thus hindered and the infant usually screamed in distress, bringing other baboons to its aid. Thus, in summary, the same male was likely to be an infant's possible father, its protector, and its exploiter.

Summary and Discussion of Social Relationships

Adult Males

Fig. 37 provides a summary pictorial representation of the close relationships between new mothers (months one through three) and adult males, condensing the data in Table 12 by assigning to each mother-male relationship a score of 1 for each entry in the table for that pair (Table 12, last column), and plotting in Fig. 37 only those relationships for which the score was greater than 1. The shorter the length of the line between a male-mother pair in the figure, the greater the association score between them. A condensation such as this can

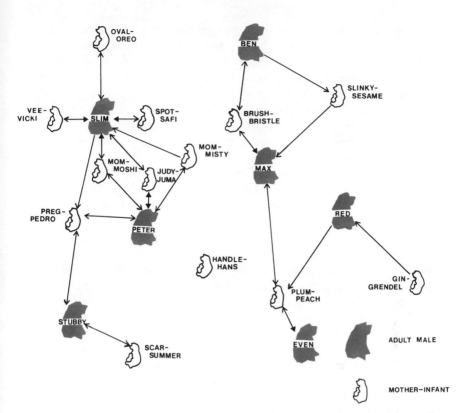

Fig. 37. *Pictorial representation of relationships between adult males and mothers during their infants' first three months of life, condensing the data in Table 12. The distance between a pair is approximately inversely proportional to their total score for all relationships (last column of Table 12); relationships are plotted only if the total was greater than 1. An open arrowhead indicates that a directed relationship occurred in the direction indicated by the arrowhead; a filled arrowhead indicates that there were two relationships (e.g., approach and grooming) in that direction.*

never do complete justice to all the data for the first three months, nor does it include data for the mother before the infant's birth or for either the mother or the infant later on (Table 13); but it does help to elucidate major patterns that exist during the early months. First, it is clear that most mother-infant dyads were associated with two males. Sec-

Fig. 36. *Adult male High Tail with infant Grendel: (top) seated in High Tail's ventrum, (bottom) riding ventrally on High Tail. In neither situation was High Tail involved in an agonistic encounter.*

ond, mothers who had one male in common usually had both their male associates in common. This is partly a consequence of only a few males being closely associated with any mothers and these males associating with the mothers of their sleeping subgroup (Fig. 17). The males that were associated with mothers usually were fully mature rather than subadult or newly adult males, and the associated males tended to come from the upper half of the dominance hierarchy. Top-ranking Slim and mid-ranking Peter were the males who were associated with the most mother-infant pairs.

These patterns of male associations can be compared in more detail with the sleeping grove subgroupings depicted in Fig. 17. Mothers were associated with the same males that were in their sleeping subgroup; more specifically, they were usually associated with the adult males who were most consistently in their subgroup. The left subgroup had fewer adult males but the same number of infants as the right subgroup, and this is consistent with the observation that Slim and Peter of the left subgroup were each associated with more mothers than were any other males. Only for Plum and Mom was sleeping grove membership (Fig. 17) contrary to the data on mating associations (Appendix 2 and Table 13) and to ad libitum data on their associations before the birth of their infants. Their male and their female associations (see below) after the birth of their infants were consistent with their previous associations, but not with their general sleeping subgroup membership.

DeVore (1963) reported that all mothers associated with a group of dominant males who prevented other animals from access to young infants. Ransom and Ransom (1971), and Seyfarth (1978), in contrast, described a variety of individual pairwise relationships, a description more consistent with the data from this study.

ADULT FEMALES

Can the associations of adult females with mothers be depicted in the same way as the male-mother associations? These female associations do not emerge with the same clarity as do the male associations. First, more females than males are associated with each mother, as would be expected from the availability of 50 percent more adult females than males (this takes into account deaths, maturations, and migrations). Second, these female associations appear to be more diffuse or of lower intensity than the male associations, most receiving total scores (as in Table 12, last column) of 1 rather than the 2 to 5 of male associations. Omitting interactive approaches and omitting relationships with a score of only 1 made little difference in relationships with males. Doing so with the relationships between mothers and other females eliminates most of the relationships. Thus no useful pictorial diagram can be made for the female-mother associations. Is this result an artifact of the particular condensation of the data? If not, what are the social factors producing this difference? First, I think this is an accurate reflection, not of female-female relationships in general, but of female-mother relationships in particular. Lower-ranking females who sought out a mother-infant pair sat near the mothers and appeared in the data as their neighbors, but did not often interact with them. In contrast, approaches by higher-ranking females were usually followed by interactions, both because the high-ranking females were more likely than were low-ranking ones to initiate interaction and because mothers usually reacted to these higher-ranking females; one of the mothers' primary reactions was avoidance. Thus there was some tendency for higher-ranking females, especially infant-pulling ones, to be frequent interactants, but not to spend much time as neighbors, and the converse tended to be true for lower-ranking females.

So the difference in dominance rank between a mother and another adult female and the presence of other higher-ranking female and male associates, as well as a female's tendency to pull infants and whether that other female herself has a young infant, all affect the degree of association between the two, modifying their tendency to associate at other times and producing a range of associations more complex than the associations with the adult males. Yet, mothers' female associates, like their male associates, came disproportionately from their sleeping subgroups. This is especially true if we consider all mothers other than Plum and Mom. For all the other mothers there was a total of 25 close relationships with other females (by the criteria used for the male relationships), 17 of which were with females in the same subgroup. Three of the remaining 8 associations that crossed subgroup lines were between females with like-aged infants. In contrast, 8 out of 9 of Mom's and Plum's associations as new mothers or with new mothers were with females of the other sleeping subgroup. However, even this inconsistency is reduced when we consider that during October 1976, Peach's first month of life, Plum was always in the "right" subgroup in the sleeping trees, the same subgroup that contained the adults with whom she was associated both during that month and before then. Thus only Mom consistently slept with one subgroup but interacted with the males and females of the other.

We have seen that those individuals who primarily compose a new infant's world are a subset of those available in the group, a subset that is different for different infants, with some consistency over various activities and with some continuity with the mother's relationships before the infant's birth. Thus although parturition is an event that brings dramatic change in a mother's social life, certain relationships are maintained or intensified at that time. The subgrouping patterns may result in a slight partitioning by genetic relatedness, but both the extent of this partitioning and the net beneficial or harmful effects of various associates require further investigation.

Summary and Discussion of Specific Forms of Interaction

The picture of the social milieu that emerges is a complex one. Some behaviors and relationships were essentially independent of the existence or age of a female's infant: the rate of overt aggression and of noninteractive approaches and the direction of dominance-subordinance relationships were all of this sort. However, several other dramatic and time-consuming behaviors became much more common soon after parturition and remained infant dependent: behaviors of fear or distress, receipt of interactive approaches, and receipt of grooming. These increased dramatically during month one of infant life and peaked during either month one or month two. The increase in these was due to other animals, not to the new mothers, who themselves maintained or decreased levels of initiation of behavior. More specifically, interactive approaches to mothers and maternal distress occurred primarily at times when their infants were in contact. Attractiveness of infants remained high until the infants were about five months of age; but apparently the reduction in mothers' feelings of vulnerability or stress was more uniformly gradual over their infants' first six months of life.

Although all females experienced these social changes after the birth of their infants, there was considerable individual variability in these experiences, either in their magnitude (Table 14) or in the identities of those who composed these mothers' social world. Some mother-infant pairs had many neighbors and others had relatively few. Some were approached and interacted with almost 30 times per 100 minutes, others about half as often. None of the major demographic or sociological variables—maternal rank, infant gender, or maternal age—proved to be clearly related to these differences, although there was a slight tendency for higher-ranking mothers as compared with lower-ranking ones to receive noninteractive approaches from more other females and interactive ones from fewer other females. This was probably due to the hesitancy of others to interact with the high-ranking mothers and the lower level of stress or fear that approaches produced in these mothers. Seyfarth (1976) and Cheney (1978) found some tend-

Table 14. Summary of neighbor relationships, rates of interactive and noninteractive approaches, and grooming time for mothers during months one through three of infant life.

Mother-infant	Average number of frequent neighbors	Interactive approaches (acts per 100 min)		Noninteractive approaches (acts per 100 min)		Average grooming time (min/100 sample min)	
		By the mother	To the mother	By the mother	To the mother	By the mother	To the mother
Brush-Bristle	6.33	3.21	22.85	.70	1.19	.390	3.13
Gin-Grendel	10.33	4.97	21.54	1.45	2.01	1.69	7.98
Handle-Hans	3.33	5.06	16.35	.56	.91	2.73	8.93
Judy-Juma	6.0	2.72	27.08	1.48	3.81	1.24	5.85
Mom-Misty	9.0	3.01	17.52	.08	.23	1.65	3.13
Mom-Moshi	6.0	2.86	24.68	2.14	5.16	1.27	7.16
Oval-Oreo	6.0	3.24	15.24	4.57	4.76	3.11	3.53
Plum-Peach	6.0	2.56	16.92	2.91	4.62	.631	5.67
Preg-Pedro	8.33	3.53	20.76	-.44	2.06	1.28	1.35
Scar-Summer	12.5	1.27	17.83	.32	.95	1.09	4.07
Slinky-Sesame	9.0	6.02	26.02	1.59	2.83	.91	7.14
Spot-Safi	10.67	6.51	19.69	2.43	4.96	4.95	5.96
Vee-Vicki	12.0	3.23	28.31	3.08	5.38	2.00	9.16

ency for higher-ranking females to be approached and groomed more than were low-ranking ones, a result that is contrary to that of the present study. I suspect that individual behavioral tendencies, that is, personality differences, sometimes override or interact with demographic or sociological effects. Unlike some high-ranking females such as Alto, the two highest-ranking new mothers in my study, Spot and Mom, were particularly aggressive individuals. Perhaps if they had not been aggressive, Spot and Mom would have been approached or groomed more.

Cheney (1978) further suggested that juvenile females seeking interaction with new mothers do so primarily with their own mothers and with female nonrelatives that are higher-ranking rather than with nonrelatives that are low-ranking. At least the latter situation does not seem to have been the case in the present study. It was not tested with Cheney and Seyfarth's own data, and additional information would be required in order to calculate the appropriate expected values based on availability in the group of higher and lower-ranking adult females with infants, in relation to each juvenile. The problem in making the test for that study is further complicated by the absence of genealogical data beyond Cheney and Seyfarth's 18-month study and therefore by their assumption of kinship relations based on behavioral association. At present the hypothesis cannot be disentangled from other interpretations.

Finally, there were two major variables in my study that were clearly correlated with maternal dominance rank, the rate of expression of fear or distress and the rate of being supplanted. By maintaining a close relationship with an adult male, low-ranking mothers can partially avoid or buffer these interactions, but they remain potent ones for mothers and their infants. For the most part, maternity exaggerated the effects of dominance rank. Occurrences of these stressful interactions may be the major factor determining the establishment of offspring dominance rank according to maternal dominance rank that has been reported for macaques (e.g., Kawai 1958, Kawamura 1958, Missakian 1972, Sade 1967) and that we have observed in this group (Hausfater et al. in preparation). Long before opportunities arise for maternal support of infants observed in the infants' own agonistic encounters (e.g., Cheney 1977), infants have had ample opportunity to learn their mothers' rank and the identity of those individuals to whom their mothers respond with fear. The infants soon seem to respond selectively to other individuals. Perhaps these early interactions account for Norikoshi's (1974) observation that by three months of age the behavior patterns of play used by each member of a pair of Japanese macaque infants reflected their future dominance relationship. Over the past seven years, in the two instances of mothers who dropped in rank (Oval, Judy) and the two in which mothers (Alto, Fluff) died when offspring were about one-and-a-half years old, the offspring assumed the ranks their mothers occupied during the period of the offspring's infancy (Hausfater et al. in preparation).

In this context recall that spatial displacements (Table 10), as well as most interactive approaches (Fig. 25) by other adult females, were made by individuals that were higher in rank than the mother. In addition, it was also primarily in response to higher-ranking females that mothers embraced or restrained their infants or directed their distress responses (Fig. 34). That is, there were few instances of undecided or reversed dominance interactions, and even nonagonistic interactive approaches followed the same partner asymmetry as did agonistic interactions. If these interactions are important in the early establishment of dominance relationships, I would predict that those higher-ranking females who were most interactive with a mother and infant would be the first to establish dominance relationships with that infant. For female infants this would mean early establishment of permanent dominance relationships; for male infants it would mean that those females would remain dominant to the infants longest—all juvenile males eventually become dominant to all adult females. The order and dynamics of establishment of dominance relationships in this group are currently under study (Hausfater et al. in preparation, Walters in press).

The increased and time-consuming social life that is imposed on mothers perhaps provides some predator protection, but it also entails increased exposure to disease and perhaps to feeding competition. Not only does it result in mothers' receiving much more grooming and thereby presumably greater rates of ectoparasite removal, but because they do less grooming of others than previously, the amount of grooming received for the amount of grooming provided is even greater than would appear at first.

However, even nonharmful interactions take time and even small amounts of time add up to a considerable total. Just 25 sequences per 100 minutes, the rate of approaches that were followed by interaction, add up to 150 sequences per ten-hour day. At even ten seconds each, these sequences alone add about 4 percent of imposed socializing time to an already tight time budget. Thus social interactions inflict at least a time cost. Some clearly have additional costs, others may be beneficial; but for most behaviors, for example, those that aid in identification of future playmates or mating partners, we can only be even more speculative about potential benefits (Kurland 1977, Rowell et al. 1968, Seyfarth 1977).

In general, mothers' behavioral interactions have traditionally been classified primarily on the basis of presence or absence of threats or overt aggression by the actor. On this basis, various authors refer to mothers' avoidance of "friendly" approaches (e.g., Rowell et al. 1968, Seyfarth 1976). Handling, grabbing, and pulling of the infant are usually considered "interest" and "investigation" rather than "aggression" or "threat." As suggested by the analyses in the present study, if we wish to understand the experiences of motherhood and infancy and to predict the consequences of these experiences, it may be more fruitful to consider interactions from the standpoint of the mothers and their infants rather than from the standpoint of their interactants. In the next several chapters I shall continue this emphasis by investigating the mother-infant relationship itself and some of the ways in which this relationship seems to be affected by the social and physical environment.

7 / Maternal Care in the Postnatal Period

THE FIRST TWO MONTHS of life—the first few weeks, in particular —constitute a period of intense dependency for the infant that produces additional stress for the mother. Although baboon infants cling to their mothers from the day of birth, most infants need some clinging assistance in the first few days (Fig. 38), some for a week or more. Individual differences in maternal response in the first weeks of life may affect the likelihood of survival of this latter group and of other high-risk infants.

Parturition

The need to keep up with the group during the day seems to be a particular strain for new mothers and may be the major advantage of nighttime births, which offer at least a few hours of rest after parturition. Most primate infants are born at night or during the very early hours of the morning (Jolly 1972; for differences in patas monkeys, see Chism et al. 1979). For savannah baboons this means that births usually occur when the group is in the sleeping trees. When we arrive on the morning of a birth, we often find an infant with its hair still wet and matted, and its mother with bloody hands, mouth, and perineum. In these cases the umbilical cord is usually still quite light in color, thick, and pliable. After a few hours mother and infant are dry and the cord is dark and stringy with little flexibility. Probably these infants were born soon before our arrival. In other cases the appearance of mothers and infants when we first see them in the morning suggests that birth occurred at least several hours earlier than that, perhaps late the previous evening or night.

Mom went into labor with Moshi the evening of 31 May 1976, reaching a stage of labor in which she alternately jerked her limbs and

Fig. 38. *Judy assisting Juma's clinging by embracing him as she fed on the afternoon of the day Juma was born. Ordinarily she would be using both hands for feeding.*

rested stretched out on her side on the ground after all other group members had ascended the sleeping trees. Finally, as it became quite dark, Mom too ascended the sleeping trees, and I left the group when the baboons could no longer be discerned. By 0500 the next morning Moshi and Mom were dry, Moshi probably having arrived the previous night.

Occasionally infants are born during the group's day journey, as were Juma and Peach during this study. Judy appeared with Juma in mid-morning after not having been seen by the observer for about an hour (S. Altmann personal communication). By the end of that day and after considerable traveling by the group, both mother and infant appeared even more fatigued than most females on the day of parturition. Plum was more fortunate; the group traveled little after Peach's mid-afternoon birth and Plum was able to rest overnight before a long journey on the following day. Plum's labor and parturition are described in more detail in Appendix 2.

Descriptions of one or more primate parturitions are gradually accumulating, primarily from captive animals of various species (see, e.g., Abegglen and Abegglen 1976, Bowdent et al. 1967, Goswell and Gartlan 1965, Gouzoules 1974, Hopf 1967, Love 1978, Nash 1974). These reports indicate considerable variability, even within species, in the extent to which mothers separate themselves from the rest of the group and in the extent to which other group members are attracted to a birth. We have observed comparable variability in the several births we have seen (in one case we observed the entire period of labor and the birth). The overall description in the literature of labor and birth corresponds fairly well to Mom's and Plum's labor.

The Neonatal Period

All baboon mothers, even primiparous mothers, seemed automatically to clutch their infants from birth and to respond to their neonates' distress cries with such "embraces." During these first few days of life the most conspicuous difference that I observed between primiparous and experienced mothers was in their subsequent responses. Rates in focal samples were too low for statistical analysis; descriptions are based on focal plus ad libitum sampling. If an infant was already in contact and being held or cradled but the infant cried or gave rooting responses in an attempt to reach the nipple, an experienced mother usually lifted the infant higher on her ventrum by means of repeated embracing movements, as a result of which the infant usually reached the nipple, clamped on with its mouth, and ceased its distress cries. Not only did the infant obtain its essential nutrition this way, but it was provided with a fifth anchor-point when riding, an advantage that seemed appreciable in the first few days of life. In contrast, primiparous mothers seldom gave repositioning or sucking aid even when, in the most extreme situations, the infant was carried upside down and backward, and required constant embracing by its mother. Infant contact seemed to be one undifferentiated state to these mothers. One often sensed that they were unresponsive to, or puzzled by, their infants' continued distress, despite the fact that nearby group members responded to the cries by watching the pair, moving closer, and increasing the repeated grunts that are given to mothers and infants (Gilmore in preparation). Most mothers learned quickly, though: increased responsiveness sometimes was observed by the end of the first day and certainly in the next few days.

Only in the case of the one totally incompetent mother, Vee, did the incompetence seem to have serious consequences. The fact that her infant, Vicki, was so poorly treated on her day of birth (see Appendix 2) could not be compensated for by the mother's appreciable improvement by the second day and relative competence later on. The early deprivation seemed to be the most likely cause of Vicki's death in the fourth week of life.

The recent research of Mason and his colleagues (e.g., Mason

Fig. 39. *Vee holding her dead infant, Vicki, and restricting juveniles' investigations of the corpse.*

1978, Mason and Berkson 1975) has emphasized the superiority of a responsive "attachment figure" to the immobile surrogate originally used in their laboratory rearing studies of rhesus monkeys. However, even the most unresponsive wild baboon mother is probably more appropriately responsive than the researchers' most sophisticated surrogate (a dog). It remains for future field and laboratory research to document more adequately the normal range of responsiveness and to determine the consequences of various levels of normal responsiveness both for survival and for such deficits in development as the cognitive deficiencies found by Mason (1978).

Mothers persist in the apparently automatic embracing of their infants even after infant death. They continue to carry the decomposing and increasingly dehydrated corpse, despite the fact that this usually means that they walk three-legged, setting the corpse down whenever they stop to feed and then retrieving it again, surely a tiring and difficult way to forage for several days. During 1969, one mother in another group often carried the corpse of her infant in her mouth or walked on all fours, dragging it on the ground. After several days the head of the corpse was gone and the corpse was quite dehydrated. On the fifth day only some dry extremities remained. As the mother sat she

Fig. 40. *An adult female of another baboon group gnawing on the dried remains of her infant's corpse, which she had carried for five days.*

began to gnaw on the hard dry piece in her hand (Fig. 40), occasionally looked at it, seeming "puzzled," and eventually stopped gnawing it, clutched the scrap again and walked off. This is the only time I have seen such behavior. After about three days other mothers leave the corpse on the ground for gradually increasing periods of time while they forage at greater distances away, eventually either lose it or leave it, looking back at the corpse with repeated signs of conflict and ambivalence and sometimes giving alarm barks. The two corpses I have recovered after desertion were both badly decomposed.

Maternal Style

By the end of the second week of life, all healthy infants climbed clumsily about in the mother's ventrum and made their first attempts to break contact. This was a period of close maternal attentiveness whenever mother and infant were not in contact and of rapid return to contact at the slightest disturbance of any sort. These first stages of separation were initiated primarily by the infant and were sometimes limited by the mother. Although most mothers were protective of their infants when intruders approached during the first months, mothers differed considerably in reaction to infant exploration during this period. The most restrictive mothers, Handle and Slinky, allowed virtually no break in contact for almost two months (Table 15, column 3). The more "laissez-faire" mothers tolerated separation, although they themselves seldom moved away from their infants during the first month. These laissez-faire mothers watched or followed their infants less often, even when the infants were quite young (Table 15, column 4).

By the end of the second month most mothers initiated some separation, and the infants oriented to and followed their mothers. During month three restrictiveness and overt maternal attention to the infant rapidly disappeared except during emergencies, whereas infant attention to and following of the mother became the norm (Table 15, column 5). This was accomplished partially through one of the few examples of what I would call teaching in baboons. A mother began to take a few steps away from her infant, paused, and looked back at the infant. As soon as the infant began to move toward her, she again moved slowly away. At first this sequence was repeated every few steps, but soon a mother seemed to be able to initiate a long bout of following by just one such pause. Hinde and Simpson have described and well illustrated this behavior for caged rhesus macaques, labeling it "mother-infant leaving game" (Hinde and Simpson 1975).

Basically, mothers could be characterized dichotomously in the

Table 15. Age of infants (in months) at transitions in mother-infant relations. Data are included for all mother-infant pairs observed at least during month one. Infants are ordered by maternal dominance rank at parturition. See Table 3 for ages at which each infant was observed.

Infant	Rating of maternal style	Mother stopped restraining infant	Mother first rejected, ignored, and stopped following infant	Mother first increased distance more often than decreased it	Infant first made at least 90% of the contacts	Mother first bit, hit, or pushed infant
♀ Safi	Laissez-faire	½	½	1½	2	2
♂ Moshi	Laissez-faire	½	1½	2	2½–3½	2–4½
♀ Misty	Laissez-faire	½	1½	2	a	a
♀ Vicki	Laissez-faire	½	0	—	b	—
♂ Pedro	Laissez-faire	1	2	1	b	1
♀ Summer	Restrictive	2½	4	2	3	5
♂ Grendel	Laissez-faire	1	1	1	2	1
♀ Sesame	Restrictive	2	2	3	4	3½
♂ Hans	Restrictive	3	5	4	5	5
♂ Peach	Restrictive	>1	>1	—	—	—
♂ Bristle	Restrictive	1½	2	2½	7	5
♂ Juma	Laissez-faire	½	½	1	2	a

a. Not before infant death at two-and-a-half months.
b. Not before infant death at eight months.

first two months as being either in the range rejecting to laissez-faire or the range protective to restrictive in their behavior toward their infants, as summarized in Table 15. In comparing the first group (hereafter called "laissez-faire" mothers) with the second (hereafter called "restrictive" ones), we find that the laissez-faire mothers not only restrained their infants less and completely stopped doing so when their infants were younger (Table 15, column 3)—the main basis on which the initial classification was made—but that they also rarely if ever followed their infants. They seldom made contact with their infants (Table 15, column 6), and at a younger infant age they increased the distance between themselves and their infants more often than they decreased it (Table 15, column 5). At an earlier age they ignored their infants (Table 15, column 4) or directed punitive behaviors toward them (Table 15, column 7) when the infants attempted contact or suckling (see also Chapter 8).

Seven of the 12 mothers for whom there were at least some observations during month one were classified as laissez-faire, 5 as restrictive. The average dominance rank at the time of parturition was 6.1 for the laissez-faire mothers, 11.2 for restrictive ones (recall that by convention the top-ranking animal is assigned the number 1, the next is number 2, and so on). Six of the 7 higher-ranking mothers were laissez-faire; 4 of the 5 lower-ranking ones were restrictive (Mann-Whitney test on the rank order of these females gave $U = 29$, $P = .05$). Three of six female infants and two of five male infants had restrictive mothers. Mom, the only mother who was observed during month one with each of two successive infants, was laissez-faire with both female Misty and male Moshi. Thus maternal rank was a good predictor of maternal style; sex of infant was not.

Is restrictiveness in some sense a reflection of general attentiveness or nervousness? The glance rate data provide suggestive evidence. For the seven 1975–76 mothers for whom I also have 1974 glance rate data, a dichotomous assignment for glance rate and for dominance rank in 1974 can be made dependent on whether an individual was above or below the median in these variables (Table 16). Then we can ask how many females were in each of the four classes; for instance, were dominant females also calm (low glance rate)? Next we ask whether being dominant or being "calm" is a predictor of being what I have called restrictive or laissez-faire as a mother.

Only four of seven females were both high-ranking and calm or low-ranking and nervous; that is, rank was not a good predictor of glance rate for this subset of females (Table 16). For five of seven, dominance predicted maternal style: that is, high-ranking females were

Table 16. Glance rates, 1974.

Female[a]	Grassland acts/min	Woodland acts/min	Average	Average rate above (+) or below (−) median	Dominance rank above (+) or below (−) median	Maternal style[b]: laissez-faire (−) or restrictive (+)
T.T.	7.82	4.92	6.37	+	+	
Alto	4.39	2.98	3.69	−	+	
Mom[b]	6.43	5.76	6.10	−	+	−
Lulu	6.80	5.07	5.94	−	+	
Preg[b]	7.18	7.78	7.48	+	+	−
Scar[b]	9.00	7.74	8.37	+	+	+
Oval[b]	6.39	5.81	6.10	−	+	
Fem	6.87	7.90	7.39	+	−	
Jane	7.13	4.90	6.02	−	−	
Slinky[b]	7.61	6.05	6.83	+	−	+
Plum[b]	8.78	9.50	9.14	+	−	+
Brush[b]	7.25	7.73	7.49	+	−	+
Este	5.87	5.70	5.79	−	−	
Judy[b]	5.05	6.86	5.96	−	−	−

a. Females are ordered by decreasing dominance rank, high to low.
b. Females whose infants were born during the 1975–76 study.

laissez-faire and low-ranking ones restrictive, as noted above for the complete set of mothers. However, six of seven females were consistent in their glance rate and maternal style: only Preg was above the median in glance rate but classified as laissez-faire as a mother (and she was the least laissez-faire of that group of mothers). Thus in this limited sample glance rate was an even better predictor of maternal style than was dominance rank, a fact suggesting that maternal style may be partially related to general temperament, but not suggesting that temperament and rank are independent. As Chance (1967) proposed, differential social attentiveness may be an important characteristic of rank structure in primate groups.

Ill health and infant death occurred at a slightly higher rate among infants of laissez-faire mothers (five out of seven versus two out of five for infants of restrictive mothers). It is not clear whether this association is a real one and, if so, what the nature and direction of causality is. As mentioned above, Mom was generally at least as laissez-faire with her weak infant, Misty, as she later was with her healthy infant, Moshi, even tolerating juveniles' carrying Misty around while Misty repeatedly bumped her head and often screamed. No general difference could be detected in the behavior of the two most restrictive mothers, Handle and Slinky, the former with a healthy infant, the other with a weak one. Thus infant health does not seem to be a major determinant of overall style of mothering, whereas dominance rank does. However, mothers did respond to worsening of infant health by increased protective behavior. This was observed with female Oval when her yearling, Ozzie, had a brief foot injury, probably an infection caused by a thorn puncture; Ozzie was quite independent by that age but when he had trouble walking Oval stayed behind and waited for him and sometimes carried him. Spot had begun rejection, or "weaning," behavior with two-and-a-half-month-old Safi when both mother and infant suffered severely from what we presumed was a virus (see Appendix 2); Spot responded with increased care to Safi's return to total dependence. Both Mom and Preg did the same when Misty's and Pedro's health deteriorated (see Appendix 2).

Differences in maternal style seemed to represent both lifetime and immediate responses of individuals to the pressures of their social world; and as shown in Chapter 6, at least the immediate pressures were a function of maternal dominance rank, expressed in agonistic interactions and infant pulling. A mother's caretaking responses, in turn, affected her infant's access to that world. For example, during their first three months, there was some tendency for infants of laissez-faire mothers to be groomed by individuals other than their mothers more than infants of restrictive mothers (Table 17). This was true regardless of maternal rank (see data for Juma and Grendel), but since maternal rank was correlated with maternal style, the overall consequence will be that in the long run infants of high-ranking females get groomed by others more than infants of low-ranking ones.

Only Rowell et al. (1968) provide quantitative descriptions of variability in maternal care and its consequences for baboons. They observed restrictiveness in low-ranking mothers and found that it disappeared when high-ranking females were removed from the cage. Since neither Rowell, DeVore (1963), nor Ransom observed restrictiveness in the wild baboons, Ransom and Rowell (1972) speculated that it might be a response to captivity in baboons, although restrictiveness has been observed in both caged and free-ranging macaque colonies (see review in Berman 1978).

Both Ransom's and Rowell's field studies of baboons were conducted on expanding groups with low death rates, groups that were probably comparable in some ways to those examined in Berman's Cayo Santiago study of rhesus monkeys and in more recent years at Robert Hinde's rhesus monkey colony at Madingley. Under these demographic conditions kin-based subgroupings can readily form

Table 17. Grooming of infants during months one through three, by their mothers and by other individuals; total time spent per 100 minutes.

Maternal style	Mother-infant	Grooming by others	Grooming by mother
Restrictive	Brush-Bristle	.045	.220
	Handle-Hans	.193	1.259
	Plum-Peach	.262	4.430
	Scar-Summer	.000	.211
	Slinky-Sesame	.279	1.097
	Average	.156	1.446
Laissez-faire	Gin-Grendel	.246	1.464
	Judy-Juma	1.404	.917
	Mom-Misty	.010	.239
	Mom-Moshi	.462	1.420
	Preg-Pedro	.028	.015
	Spot-Safi	.596	1.525
	Vee-Vicki	.880	2.550
	Average	.518	1.161

(S. Altmann and J. Altmann 1979). Berman (1978) recently compared changes in the data from Hinde's colony. She indicated that small kin groups have developed within enclosures in the years since formation of this colony and speculated that this may have resulted in maternal and infant behavior in the captive group increasingly resembling that of the kin-organized rhesus from the free-ranging Cayo Santiago colony. She suggested that mothers are less fearful and restrictive when they can surround themselves with close kin.

The usual conditions of captivity may exacerbate social stress by constraining or at least changing the options available to mothers, for example, by limiting space and providing only one adult male with a small group of females. However, social as well as ecological stress also occurs in natural habitats and apparently influences maternal style. The extent and nature of the influence probably depends on each individual's associations and options and on her ability to capitalize on those options.

Differences in maternal style within and between groups probably affect the rate at which infants develop independence from their mothers, as will be shown in Chapter 8. A number of developmental differences may then be a direct consequence of maternal style or of differences in rate of independence that may be caused by style of mothering.*

---

* Initial observations are available for five of the 1975–76 juveniles and four of the 1975–76 mothers who gave birth in the latter part of 1979. Gin and Spot are again very early rejectors, Preg intermediate, and Plum late. Also a late rejector is Spot's primiparous sister Dotty, despite her high dominance rank. For primiparas Cete and Fanny it is too early to determine more than that they are not very early rejectors, consistent with what I would predict from both their dominance rank and the maternal behavior of their own mothers. With respect to the effects of a juvenile's experience with infants on subsequent maternal care, Cete, who interacted little with infants, as well as Dotty and especially Striper, who interacted more, all appear to be quite competent with their first offspring. Particularly striking was Cete's compensation for her infant's initial difficulty in clinging. Nazu, however, who as a juvenile interacted often with infants, was not very attentive to her first infant. She often sat partially on it, and she showed little interest in it when it was kidnapped by Summer on its sixth day of life. The infant died three days later, and Nazu carried the corpse for two days thereafter (J. S. Dillon, personal communication). Thus the preliminary evidence provides some support for the idea of intra- and inter-generational consistency of maternal style, but no pattern emerges in the relationship between a juvenile's experiences with and interest in infants and her subsequent competence with her own first infant.

8 / Infant Development and Mother-Infant Spatial Relationships

WE ARE NOW READY to turn to the infant itself, its development and the progress of its independence from its mother during the first year of life. Again, I shall consider individual differences as well as those features common to all infants. I shall examine the effects of ecological variables, maternal time budgets, and maternal style on mother-infant relations and independence. First, though, I shall consider the general pattern of infant maturation.

Physical Maturation

Baboon infants change appreciably in both appearance and behavior over the first year of life (see also Ransom and Rowell 1972). At birth the infant has skin that appears bright pink and a natal coat of black hair along with disproportionately prominent ears, and the males have a particularly visible pink penis. This coloration contrasts strikingly with the adults' yellow-brown coat and grayish skin.

During the third month of life healthy infants all developed some gray skin pigmentation in their hands and feet and some gray mottling in their paracallosal skin, as well as some gold hair in their coat, particularly around the wrists and brow region (Fig. 33). By month four Sesame and Pedro were noticeable for (1) absence of gray skin coloration, (2) loss of hair on tail, and (3) absence of gold hair. Their skin color change occurred later than did other infants, and not only was their coat color change delayed, but the hair was thin and white or beige when it did change (Fig. 42). These abnormalities may be related to nutritional deficiencies (S. Altmann in preparation), and it is notable that they became obvious at about the age that caged baboon infants whose mothers were fed a low-protein diet dropped below the range of weights for infants of mothers on normal laboratory diets (Buss and Reed 1970).

Fig. 41. *Three playing infants. Five-month-old Eno* (left) *has lost much of her black natal coat but still has pink ears. One-month-old male Bristle* (crouched) *has the complete natal coat and skin, including poor definition between callosities and paracallosal skin* (toward camera, center). *At the age of nine months, male Ozzie's coat, paracallosal skin, and ears have completely lost their natal appearance. Bristle's mother, Brush, is partially visible in the background.*

By six months of age infants have reached a number of very visible milestones (DeVore 1963, Ransom and Rowell 1972). Virtually all of their natal coat and most natal skin color have been replaced by skin and hair similar in appearance to that of adults (J. Altmann et al. 1977, K. Rasmussen in preparation). However, a few tinges of pink remain in the muzzle and ears almost until the first birthday, and the male's scrotum remains pink for at least another year after that.

At birth infants' locomotory abilities are quite undeveloped (see Hines 1942 for macaques, Rose 1977 for anubis baboons). During the first weeks objects that baboon infants put in the mouth are picked up directly with the mouth rather than with the hands. Their hind limbs are even more flexed than their forelimbs and their gait is extremely wobbly during the first few weeks. Locomotory attempts usually end with the infant falling over after only a few steps. When not clinging to their mothers, neonates sit hunched over or flop on the ground. It also seems to take a few weeks before the infant recognizes its mother and

Fig. 42. *Pedro after his coat had turned white.*

Fig. 43. *While his mother Gin is being groomed, neonate Grendel explores his immediate surroundings. A juvenile investigates the infant with the common flexed-arm pose (Appendix 4), which is often accompanied by lipsmacking and soft grunts.*

selectively orients and clings to her in preference to other animals (see also Horwich and Manski 1975, Sugiyama 1965, Wooldridge 1971, indicating a similar timing in langurs and colobus monkeys). This lack of selectivity may make infant snatching especially dangerous in these early weeks: at this stage infants cling tenaciously to whoever carries them.

Initially, infants ride only ventrally. Soon, however, some attempt to climb up the mother's legs to her back. Fred, Alice, Ozzie, Eno, and Pooh all rode dorsally, although Pooh had considerable difficulty doing so until she was about eight months old. Fred, Ozzie, and Alice all did so with ease by three months of age (Jane Scott, personal communication), as did Eno. Summer, Grendel, Sesame, Juma, Safi, and Moshi all rode dorsally by two months of age, Oreo a little later. However, Pedro, Bristle, and Hans never did ride dorsally, nor did Misty by the time of her death at a little over two months. We have less systematically observed this variability in dorsal riding in other Amboseli baboon infants. It does not seem to be a useful age or developmental marker, as was indicated at first (DeVore 1963). Nor is any strong gender difference apparent.

Physical and behavioral maturation proceed at different rates for

Fig. 44. *Summer riding on her mother's back. After she was about six weeks old, Summer did most of her riding this way and soon became particularly adept at the "jockey" form of dorsal riding.*

Fig. 45. *While his mother is being groomed, Grendel, still less than a month old, climbs about on a small log. Perhaps because he was allowed to explore from his first days of life, he developed considerable locomotor competence quite early.*

various characteristics. Infants became quite competent in locomotion on the ground by the end of the first month of life, and during month two they mastered climbing over small logs and other obstacles. By this age, too, they were able to manipulate the plants and other objects they found on the ground, no longer getting these to their mouth just by mouthing them.

During the third and fourth months of life, infants began to spend increasing amounts of time in peer interactions, particularly play (Fig. 46; Owens 1975), though most did not stray far from their mothers to do so. At this age they began to feed on particularly accessible foods (see Chapter 5), could negotiate fairly well within trees, and could climb into and out of some umbrella trees but could not climb any of the fever trees, perhaps because of their smooth bark. They were still dependent on their mothers for virtually all food and transportation, but increasingly it was they who attended to their mothers, who maintained contact and proximity, rather than vice versa (see Chapter 7 and below).

For the second six months of life, my data for behavior outside the mother-infant pair come primarily from ad libitum observations,

Fig. 46. *Several infants* (left to right: *Ozzie, Bristle, Eno) play near the youngest's mother, Brush.*

checks of physical maturation, and the focal samples of mothers (an infant's own mother and others). A more complete picture of infants' interactions with others must await the more detailed studies of the older infants themselves (S. Altmann in preparation, N. Nicolson in progress). By six months of age, umbrella trees were negotiated with ease as were the lower and more branched of the fever trees. By eight months infants descended alone from even the tall, unbranched trunks of the fever trees used as sleeping trees.

During months 9 through 12 dominance relationships began to be evident among the infant dyads. These older infants also lost their general immunity from spatial displacement and other aggression by juveniles, and they could not climb on or sit immediately next to many adults with impunity. Older infants still slept with their mothers but usually had to descend the sleeping trees alone in the morning. By its first birthday every infant was quite independent and perhaps could have survived its mother's death (although three infants 17, 20, and 21 months old were the youngest known surviving orphans we have had). The mothers of approximately half the infants in Amboseli resumed estrus cycles when the infants were 8 to 12 months old, most others a few months afterward; mothers then became pregnant during the offspring's second year of life. (J. Altmann et al. 1977, 1978). However, those young juveniles whose mothers had not yet become pregnant retained occasional daytime nipple contact and continued to sleep at night in the mother's ventrum and probably on the nipple, the oldest observed individual being a two-year-old male, Jake.

Despite its increasing independence, an infant's primary relationship throughout the first year is with its mother. I shall now examine in more detail the development of mother-infant spatial relationships through the first year and the changing contingencies between the behavior of mother and infant; then rejection and weaning will be considered in Chapter 9.

Mother-Infant Contact Time

Contact between mother and infant and the maintenance of mother-infant spatial proximity provide the core of my examination of the mother-infant relationship and the development of independence. The overriding importance of body contact in the healthy development of infant monkeys was dramatically demonstrated and expressively described by Harry Harlow and his coworkers over 15 years ago (e.g., Harlow 1958, Harlow and Harlow 1965, Mason et al. 1968). In the years since then it has been shown, not too surprisingly, that a stationary furry surrogate is not a totally sufficient rearing associate (see Mason 1978, Mason and Berkson 1975), and in an ingenious series of studies various workers, especially Mason and colleagues, have sought a more complete, minimally sufficient rearing partner for infant monkeys. However, the importance of contact remains; and because of this, as well as because of the ease of defining and measuring contact and making comparisons across studies, mother-infant contact and the maintenance of contact persist as major parameters of interest in developmental studies. Those who take a cross-cultural view of human infancy have also addressed the importance of how much time an infant is carried, and by whom, as well as the position of carrying with respect to nipple access, potential for interaction, and so on (e.g., Freedman 1974, Konner 1976).

For baboons in the wild, as for humans of many societies, mother-infant contact is not an exclusive activity nor one that provides just psychological comfort and nutrition for the infant. Contact is also the means whereby an infant obtains transportation, protection from predators and conspecifics, and probably protection from extremes of weather—warmth during rain or cool nights, shelter from the glaring midday sun (Dunbar in preparation, Stelzner in progress).

Contact always occurs during one of a mother's ongoing activities and, as seen in Chapter 5, during most of the day mothers are feeding or walking. One question, then, is how a mother's activities are affected by infant contact and how in turn contact is affected by the other demands on a mother's time and attention.

The time that infants spent in contact with their mothers was calculated from the continuous record by using CRESCAT to identify onsets and terminations of contact, to calculate bout durations, to cumulate actual time in contact, and to cumulate total sample time-in-sight. The results, with each mother's data pooled for a month in a single point, as before, are presented in Fig. 47. In addition, a linear regression was fitted to the complete set of daily data points, and a plot of the complete set of residuals is provided in Appendix 5, along with plots of the residuals from the linear regression for each infant separately, clearly showing the tendency for individual infants to be consistently more (or less) independent than others. Despite the fact that the values are percentage points, for purely visual display a linear model provides a sufficiently good fit to the data with well-distributed residuals. As would be expected, the fit is poorer in the first month and after month eight, but transformations did not improve the overall fit of the data, probably because the nonlinear portions were very short. In particular, departures from linearity are due to infants' being in contact

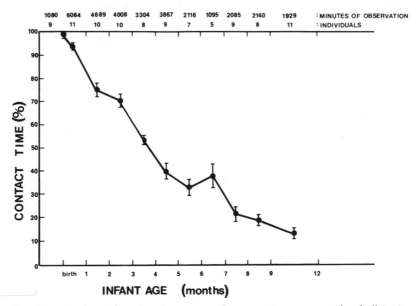

Fig. 47. *Mother-infant time in contact, from continuous records of all initiations and breaks of contact. Means and standard errors of the mean were calculated as indicated in Fig. 21.*

more than predicted during the first month; and the age at which contact time reaches zero depends on when an infant's mother conceives her next infant.

In Fig. 48 daily contact data were first pooled by month for each infant and then averaged separately for infants of restrictive mothers and for infants of laissez-faire mothers. The upper part of Fig. 48 includes all data, even those from periods in which infant ill-health was appreciable. Even so, it is clear that infants of restrictive mothers spent more time in contact throughout the first half-year of life than did infants of laissez-faire mothers. For each month, their contact time was about the same as that for the other group one month earlier. The difference is more striking when data from periods of ill-health are excluded, as shown in the lower part of the figure: that is, illness tended to increase contact time, particularly during the first few months of life.

A similar analysis of contact time was made comparing male and female infants within each type of mothering. Most sickness occurred among female infants of laissez-faire mothers, resulting in somewhat greater contact time (6 percent more contact per month) for female infants as compared with male infants of laissez-faire mothers. However, for infants of restrictive mothers, differences were small and were reversed in direction from month to month. When data from months of illness were removed, differences were small and reversed in direction from month to month both for infants of laissez-faire mothers and for infants of restrictive mothers. Thus in this study there were no differences in contact time attributable to sex of infant.

Infant's Use of Space

The young infant's world can be viewed as a circle with its mother at the center. We might, then, ask what percentage of its time does an infant of each age spend within circles of different radii? Because I do not have continuous records for distances greater than arm's reach, I used the point-sample records, estimating the probability of an infant being at various distances from its mother by the number of point samples at that distance divided by the total number of point samples. In Fig. 49 monthly percentages were calculated for each infant and were then averaged over all infants. In Fig. 50 related data through month eight are presented in such a way as to highlight individual differences. The expansion of the infant's physical world is abundantly clear from these data.

Over the first six months of life, contact time dropped steadily from 100 percent in the first two weeks to about 32 percent in the sixth month. After the first two weeks, time spent out of contact but within

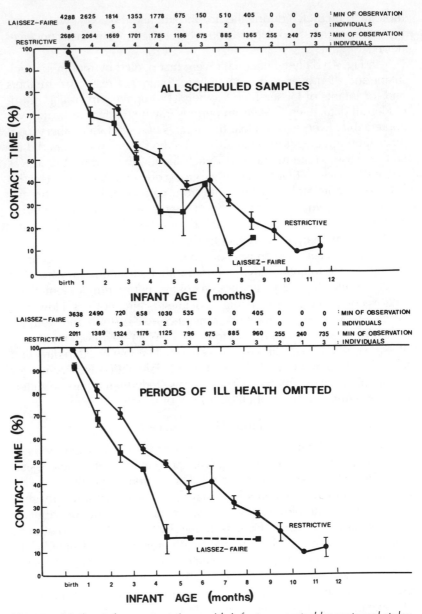

Fig. 48. *Mother-infant contact time, with infants separated by maternal style:*
(top) all months for all infants who were observed at least during month one;
(bottom) months of illness omitted. Means and standard errors of means were
calculated as indicated in Fig. 21.

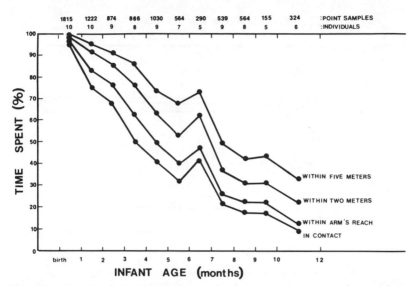

Fig. 49. *Distance between mother and infant at various ages, derived from point samples. Each line indicates the cumulative percentage of time spent at that distance or closer; the distance, parallel to the y axis, between any two lines is equal to the percentage of time spent between those distances.*

arm's reach remained at about 8 to 10 percent of total time (but it was a decreasing proportion of the noncontact time). Time spent at each of the greater distances showed successive increases. By the sixth month, approximately a third (32 percent) of the time was spent more than 5 meters away, another third (36 percent) out of contact but within 5 meters; whereas during month two, only 5 percent of the time was spent more than 5 meters away, another 20 percent out of contact but within 5 meters.

Although infants made increasing use of space more distant from their mothers, their use of this space was not at all independent of the position of their mothers. Employing a model of mother-infant distance based only on available area and taking 1 meter as a mother's arm's reach, one would predict that infants would spend approximately three times as much time in the zone between arm's reach and 2 meters ($4\pi - 1\pi = 3\pi$) as in the zone between contact and arm's reach ($1\pi = 1\pi$), seven times as much between 2 and 5 meters ($25\pi - 4\pi = 21\pi$) as between arm's reach and 2 meters ($4\pi - 1\pi = 3\pi$). Yet at no time during the first year did the actual distribution even approach the random one (Fig. 49). Always, infants used space nearer their mothers appreciably more than expected.

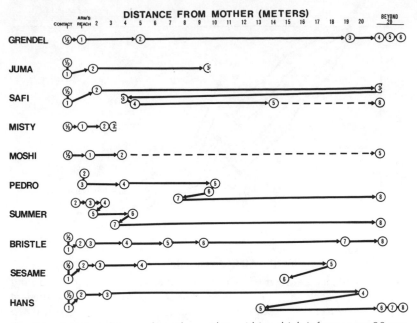

Fig. 50. *Smallest distance from the mother within which infants spent 90 percent of their daytime at various ages; infants are ordered by approximate degree of maternal restrictiveness, with the least restricted infant at the top. The half-circles for Juma and Misty indicate their deaths in the beginning of the third month; the two half-circles for Safi separate month three into periods before and after the onset of illness (see Appendix 2).*

Examination of Fig. 50 reveals the considerable variability among the infants. In this figure I took as a criterion the distance from the mother that accounted for 90 percent of an infant's daytime. For example, during month two, a radius of 2 meters from the mother on the average accounted for 90 percent of an infant's daytime, but for Grendel the radius required was 5 meters, for Moshi and Misty (Mom's two successive infants) it was between 2 and 5 meters, 2 meters for both Juma and Safi, arm's reach for Pedro. These are all the infants of laissez-faire mothers. By contrast, all the infants of restrictive mothers were within arm's reach 90 percent of the time during month two.

By the end of month eight all infants spent at least 40 percent of their time more than 5 meters from their mothers. By month six for most infants, and by month eight for all, the distance from their mothers required to encompass 90 percent of their time was no longer a useful measure—it was over 20 meters, and I often could not locate the two at the same time.

During the second half of the first year of life, contact time declined from 30 percent at month six to 10 percent by about month ten. Younger members of this class rode during part of long day journeys. Otherwise, contact at this age occurred primarily during external alarms and during a long, dozing nursing bout, commonly in the early evening, at times when mothers were either resting or involved in grooming interactions with others (see "Infant Contact as a Contingent Behavior," below).

Spatial Relations: Dynamics

During the focal samples, I recorded the time and the identity of the individual who effected (performed) each change in the mother-infant spatial relations (classified as contact, arm's reach, greater than arm's reach) whenever a change occurred from one of these states to another and that same actor did not immediately (within 0.05 min) again change state; if the actor did immediately change state again, the 0.05 minute and actor identity criteria were again applied. For these spatial-state records four categories of actors were distinguished: focal mother, her infant, mother and/or infant but I could not tell which, and "other" (with the individual identified). Over 90 percent (usually well over 95 percent) of all transitions were effected by exactly one member of the mother-infant dyad. Transitions for which I could not distinguish between the mother and infant as actor accounted for less than 1 percent of the transitions for most months, with higher percentages occasionally occurring at times of group alarm or during months of highest rates of state changes. During the first two months of infant life, "others" effected 2 to 5 percent of the transitions for infants of laissez-faire mothers by carrying the infant. For restrictive mothers total rates of transition were too low during these early months to permit reliable estimation.

On average, 70 to 80 percent of all changes in spatial state that were made by one member of the mother-infant dyad were made by the infant. Infants made over 85 percent of the breaks in contact during the first five months, 60 to 80 percent thereafter. They made 50 percent of the contacts during month one, over 90 percent after month three. Again, however, the pattern of individual differences (Table 15, column 6) reveals that Grendel, Juma, and Safi were making 90 percent of the contacts by month two; Bristle, Hans, and Sesame not until month four or later. Data for Misty are quite variable, probably because of her poor health. Moshi made 61 percent of the contacts in month two, 99 percent in month five; no data are available for him for months three and four.

Data on changes from one of the three spatial states to another were then reanalyzed by collapsing the categories into increases or decreases in the distance between mother and infant: both a change from contact to within arm's reach or to greater than arm's reach and a change from within arm's reach to greater than arm's reach were considered "increases"; changes in the opposite direction were considered "decreases." In Fig. 51 the percentage of decreases and the percentage of increases that are due to the infants are graphed for the first year of life. These provide estimates of the conditional probabilities that a decrease (and increase) will be effected by the infant, given that it was performed by only one member of the mother-infant dyad. These are approximately the %A and %L of Hinde and his coworkers, but these workers did not include transitions between contact and "within arm's reach" in their definitions of "approaches" (A) and "leaves" (L) (see Hinde and Spencer-Booth 1967, Hinde and Atkinson 1970). In a field situation, with the animals usually moving along while foraging, I found it important to include these classes—it appeared that contact and arm's reach were quite different to the animals. However, for comparison with the other studies I also reanalyze these data by Hinde's "approaches" and "leaves" in a later section.

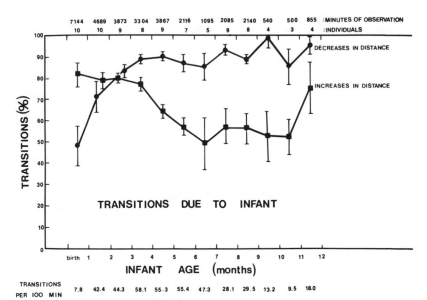

Fig. 51. *Percentage of the increases and decreases in distance between mother and infant that were performed by the infant at various ages (see text for details). The average rate of transitions (acts per 100 minutes) at each age is indicated below the horizontal axis.*

Fig. 52. *Semi-independent Moshi (top) and Bristle (bottom) attempt to ride as their mothers walk and ignore them. If a mother walks slowly the infant commonly either grabs her side for several steps (above) or repeatedly runs around in front of her and reaches out (bottom).*

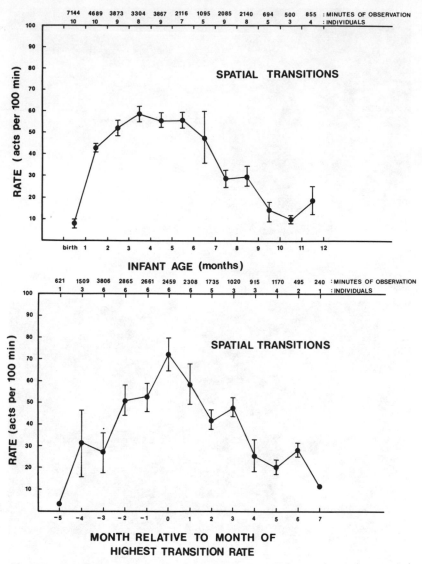

Fig. 53. *Total transition rate* (acts per 100 minutes) *for mother-infant spatial relations:* (top) *aligned by infant age;* (bottom) *aligned by the month of greatest rate for each infant, indicated as month zero on the horizontal axis.*

As noted earlier, the first stages of independence were characterized by infants primarily increasing the distance between the two and mothers primarily decreasing it; that is, the earliest stages of spatial independence were infant initiated. Only later did mothers reverse the dynamics of the spatial relationships, increasing the distance between themselves and their infants more than decreasing it (see Chapter 7). Table 15, column 5, reveals that this stage was reached later for the restrictive mothers than for the laissez-faire ones. By this stage mothers had stopped following their infants; rarely if ever paused, looked at, or waited for them; and usually ignored their distress cries as they (the mothers) moved away (Table 15, column 4).

Clearly, not only the identity of the member of the pair making the transition but also the transition rate changed appreciably (Figs. 51, 53) as the infant matured. In particular, the peak transition rates occurred soon after the time that infants switched their role and began decreasing distance to their mother more than increasing it. It is at this age that the observer has the impression that infants are frequently following or chasing after their mothers, sometimes even screeching and "throwing tantrums" (Fig. 54). I shall return to discussion of this "weaning" stage in the next chapter.

It is interesting to note that not only did the laissez-faire mothers leave their infants more than did the restrictive mothers, but their infants also left them more often, owing at least partly to the absence of restraint. Thus very different interactional patterns and rates of independence were established within dyads during the first month of infant life. Infants as well as mothers surely contributed to this interactional pattern, and most changes in spatial relationships were actually effected by the infant; but the data from this study suggest that differences between mothers were the major or initial determinant of differences between pairs.

Because infants differ in the age at which transition rates peak, the averaged data (Fig. 53, top) produce a flatter curve for monthly transition rates than do the data for individual infants. The pooled data thereby mask the appreciable developmental change in transition rates. This is illustrated by taking only the data for those six infants (three male, three female) for whom there are data from a sufficient number of continuous months to identify the peak month of transition rates. Peak transition rates occurred during month four for Safi, Sesame, and Grendel; month five for Hans; and month six for Summer and Bristle. Aligning the data for these infants by these peak months rather than by age of infant, we obtain the curve shown at the bottom of Fig. 53, with a mean of 72 transitions per 100 minutes for the peak

a.

b.

Fig. 54. *Five-month-old Safi "throwing a tantrum" as her mother apparently ignores her and continues to forage. Safi is shown alternatively chasing her mother while screeching ("eee"), grimacing, cackling ("ikk"), waving her tail in the air (a and c); and throwing herself on the ground, giving the pursed lip "coo" (b and d).*

c.

d.

month. The time course of each infant's development probably corresponds more to this pattern than to the age-determined one graphed at the top of Fig. 53. I do not know whether it is a coincidence that the peak transition rates for the three female infants were all higher any of those for the three males.

By month 7, infants were doing about 90 percent of the reductions in distance and 50 percent of the increases in distance between mother and infant. That is, when mother or infant moved closer to the other, nine times out of ten it was the infant who did so, whereas when they moved apart, the mother was as likely as the infant to do the moving. No clear change in these percentages emerged after this time, with decreases in distance due to the infant fluctuating slightly around 90 percent and increases in distance fluctuating somewhat more about 50 percent. These fluctuations are probably due to the small sample sizes for these later months: not only did I sample for fewer days per month and have fewer animals for months 10 to 12, but rates of transition dropped sharply after month 7, remaining below 30 transitions per 100 minutes throughout the rest of year one. Thus it is the rate of transitions, or changes, in mother-infant proximity rather than who effected these changes that best characterizes the dynamics of development within the second half of the infant's first year.

Infant Contact as a Contingent Behavior

In the previous section I treated an infant's contact with its mother as a state that was independent of the mother's other activities. In fact, I found that the probability that an infant would be in contact was dependent on a mother's current activity, and this contingency itself changed appreciably during the infant's first year of life. In Fig. 55 the probability of contact is plotted given two different maternal activity states: grooming or resting, and walking. Young infants, up to about four months of age, were disproportionately out of contact when their mothers were resting or engaged in grooming interactions and disproportionately in contact when their mothers were walking. For infants over seven months of age, the relationship was just the opposite: infants were much more likely to be in contact when their mothers were resting or grooming that when their mothers were walking. Another way of viewing this reversed contingency is to examine the probabilities that a mother was engaged in various activities, given that her infant was in contact. For mothers of older infants the probability was over 90 percent that the mother was resting given that her infant was in contact at the time, this despite the fact that mothers spent only about 25 percent of their total daytime resting or in grooming interactions.

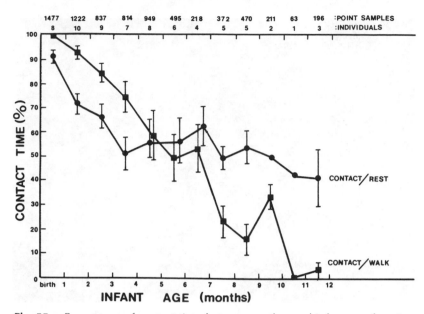

Fig. 55. *Percentage of contact time between mother and infant as a function of the mother's activity, walking in one case, resting or in grooming interactions in the other, at various infant ages.*

Fig. 56. *Young Misty has moved outside her mother's ventrum and begins to break contact with Mom as Mom grooms Misty's two-year-old sister, Striper.*

Fig. 57. *Yearling Bristle* (top) *suckles while his mother dozes; Alice* (bottom) *suckles while Alto is being groomed by her daughter Dotty.*

The data on contact, when averaged for all infants, mask the individual variability and damp the maturational effects, as was true for the change in rates of spatial transitions discussed in the previous section. Again, this occurs because the form of the graph is essentially the same for each infant, that is, each infant goes through the same stages, but the time course varies: the reversal of contingencies occurs at different ages for different infants and, in particular, occurs later for those infants whose mothers were described above as being restrictive, as shown in Fig. 58, where the data are partitioned by maternal style.

Having identified a contingency between a mother's immediate activity and the probability that her infant was in contact, we might then ask whether at some ages a mother's time budget for the day determines the amount of infant contact that day: that is, whether there are conditions under which a mother's activities are a limiting factor in determining infant contact.

Ecologists consider a resource to be limiting for some process if increased abundance of that resource results in growth or increase of the process (see McNaughton and Wolf 1973). The fact that an organism utilizes a resource or that organisms preferentially seek the resource or clump around it does not mean that the organism would be more abundant if only more of the resource were available. In the case of maternal contact as a resource for an infant, the fact that at some age infants are disproportionately in contact when their mothers are resting does not tell us whether infants would be in contact more at times of the year or in locales (e.g., zoos) in which mothers spent more time resting, or conversely whether differences found between situations or habitats can be accounted for by differences in time spent in the various maternal activity states in the several situations. Even though an infant's contact with its mother is contingent on its mother's immediate activity, that activity may or may not be a factor limiting or determining the amount of contact an infant has with its mother during a day: each day there may always be enough resting time in total so that available resting time does not limit the total amount of mother-infant contact time for the day, even for older infants. When a very young infant has been out of contact with its mother, it resumes contact virtually every time its mother moves more than a few steps. At first this happens by the mother scooping the infant to her ventrum, later by the infant itself joining its mother, as discussed previously. By the infant's third or fourth month, it uses its mother for transportation primarily if the group progression is quite rapid or the journey has been very long that day. Thus we would predict that a very young infant would be in contact more on days when its mother walked more and a somewhat

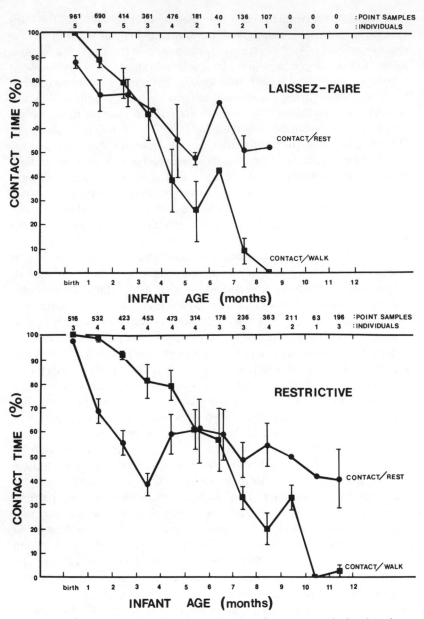

Fig. 58. *Infant contact as a function of maternal activity, with the data from Fig. 55 plotted separately for infants of laissez-faire mothers* (top) *and infants of restrictive mothers* (bottom).

older one would be more influenced by the length of the day journey. An infant more than about seven months old would be expected to spend disproportionately little time in contact on days when its mother walked a lot.

For each sample day on each infant, I calculated from the continuous records the percentage of time that the infant spent in contact with its mother (as described previously). For most infant ages this value is very close to that estimated from the much more easily obtained point, or instantaneous, samples (J. Altmann 1974, Simpson and Simpson 1977). From the point-sample data I estimated the percentage of time the mother spent in each activity that day; from the demographic data we know the age of the infant; and the length of the day journey was calculated from the day journey maps (S. Altmann and J. Altmann 1970). Using SPSS Scattergram (Nie et al. 1975), I then examined the effect of day journey length and time budget on the amount of time that infants of each age (by months) were in contact with their mothers, using as the dependent variable the contact residual, that is, the deviation in observed contact time for that day from the contact time predicted for that age (by day) from the linear regression.

When infants were in the first month of life, they were in contact so much of the time that existing variability in time budget or day journey did not affect the time spent in contact. Likewise, for infants over seven months of age, who were, overall, in contact less than 15 percent of the time, there was only 1 of 15 regressions that was "significant" at .05 level—about what would be expected by chance. However, during intermediate ages daily contact time was significantly related to maternal time budget (Table 18). Infants were in con-

Table 18. Months for which the percentage of mothers' time spent walking or the distance traveled was significantly related to the percentage of time the infant was in contact, determined by linear regression.

Month of infant life	Variable	Variance accounted for (r^2)	Statistical significance
2	Percentage of time spent walking	.17	.005
3	Percentage of time spent walking	.13	.02
4	Distance traveled	.14	.03
	Percentage of time spent walking	.10	.05
6	Distance traveled	.17	.03

tact more on days of long day journeys or when their mothers spent more time walking; they spent less time in contact on days when their mothers spent more time feeding or resting. Day journey length or time spent walking usually explained about 15 percent of the daily variability in contact residuals during these months. We can tentatively conclude that maternal time budget is to some extent a limiting factor determining the degree of independence at intermediate ages.

For young infants these results are as expected at ages when infants were disproportionately in contact while their mothers walked: the time a mother spent walking during a day was a partial determinant of contact, and therefore noncontact, time for that day. At ages for which infants were neither disproportionately in or out of contact during walking, it was day journey length that became important. Surprisingly, the expected negative relationship between time spent walking and mother-infant contact did not occur for older infants, and it is not just that the correlation failed to reach significance—a fact that could be considerably influenced by sample sizes for those months. The results even fluctuated in direction for these older infants. A finer-grained analysis, involving more infants, will be needed to elucidate the factors affecting this period, but at this time it appears that mothers' activity budgets do not limit the contact time of older infants, but do limit that of younger ones.

From a causal standpoint I would like to know whether mothers alter their behavior as a result of their infants being in contact or whether infants change to (or from) contact with their mothers as a result of the mothers' activity. Ideally, I would need to have continuous rather than point-sample data of the maternal activities as well as of infant contact so that I could examine the lag time to a transition in infant behavior given a transition in maternal activity and compare it to the overall lag time in such transitions. If it is infants who are adjusting their behavior to mothers, the contingent lag times would be shorter after a transition to a contact-stimulating activity than it is in general. It would be easy to do such an analysis by using CRESCAT, but I was not able to collect the continuous records for most maternal activities; and I therefore must be content for the time being to consider the suggestive evidence from ad libitum data and from rare events. These data suggest that the accommodation is primarily by the infants rather than by their mothers. An infant repeatedly hopping on and off its mother's back interferes with the mother's feeding and walking, especially if the infant tries to ride ventrally rather than dorsally. I often saw mothers of 4- to 5-month-olds rebuff contact attempts made while the mothers foraged. Sometimes, after being rebuffed the infant, either immediately

Fig. 59. *Grendel, screeching, temporarily regains contact with his mother after she shook him off her back and rapidly walked away from him.*

or after it "threw a tantrum," rode quietly on its mother and the mother no longer pushed it off. In similar instances, mothers of older (6- to 8-month-old) infants dislodged their infants while walking. The dislodged infant then followed its mother, sometimes throwing a tantrum, and the mother eventually sat and then immediately embraced the infant to contact and a suckling position. In these instances, the persistence of the infants did seem to shorten the time to a change in maternal activity, but the change had to occur before the infants were allowed to make contact. More dramatic cases occurred with 8- to 12-month-olds. Sometimes I was following a mother during a sample and I could not locate her infant at all. The mother began a grooming interaction or resting and immediately the infant would dart toward her from an activity over 20 meters away, make contact, and begin suckling.

Summary and Discussion

A baboon infant begins its first year of life with a distinct natal coat, undeveloped motor abilities, and a total dependence on its mother. By the end of that year, its appearance is distinguishable from that of adults primarily by size, it can perform most adult behavioral patterns, and it probably could survive its mother's death. However, it still occasionally goes to its mother in times of alarm and for rare nursing bouts and still sleeps in her ventrum, perhaps suckling, in the trees at night.

The role of the natal coat is of interest at various stages of maturation. This conspicuous coat probably does not attract predators as long as the infant spends most of its time in its mother's ventrum. Once the infant rides dorsally or runs about independently the less conspicuous brown-yellow adult pelage is probably safer. Another speculation is possible regarding the change to adult pelage. It has often been suggested that the function of a conspicuous neonatal coat is to clearly mark the infants for more tolerance and protective treatment by conspecifics. Yet, if this coat is one of the major stimuli that attracts other group members, as it seems to be, and if the attraction places an additional burden on mothers, as I indicated in Chapter 6, the change to adult pelage may be of value for mothers by reducing social pressures at a time when nutritional and ecological pressures are becoming extreme (Chapter 5). Proposals for both the advantages (benefits) and disadvantages (costs) of infant pelage remain speculative.

Contact time between mother and infant, which was 100 percent just after birth, declined at a rate of about 8–9 percent per month over this first year, with infants of restrictive mothers consistently spending more time in contact with their mothers than did their peers at the same age. Illness also resulted in greater contact between mother and infant. However, infant gender did not affect the amount of time an infant spent in contact with its mother. Although maturing infants increasingly spent time at greater distances form their mothers, at all ages they disproportionately used areas close to their mothers.

A few studies report data with unbiased estimates of time that maturing infants spend in contact with their mothers during the day. They include Struhsaker's for vervets in Amboseli (Struhsaker 1971), Berman's for the provisioned rhesus colony of Cayo Santiago (Berman 1978), and Konner's study of !Kung in the Kalahari. Data from these reports are plotted along with those from the present study in Fig. 60. In some cases it was necessary to make extrapolations or to read points from graphs. Also for the !Kung data there was a sex difference in the latter part of the first year of life, but since contact with those other than the mother showed an opposite sex difference, I just plotted the average of the data for boys and girls. The curves are all quite similar (see also Konner 1976), but the more slowly maturing baboons spend more time than the vervets or the rhesus, and the human children spend much greater time in contact than all the others, especially in the second six months. As Konner points out, the limited data for European

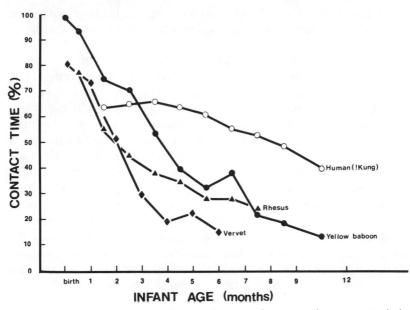

Fig. 60. *Mother-infant time in contact for several species of primates, includ-*
ing humans. Data for yellow baboons are the same as those given in Fig. 47.
These data are plotted along with available data on vervet monkeys (Struhsaker
1971), rhesus monkeys (Berman 1978), and human hunter-gatherers (Konner
1976).

and American children, both home- and institution-reared, indicate
values of mother-infant contact that are much lower than any of these,
usually less than 25 percent of the time in contact even for quite young
infants (Konner 1976).

At every infant age changes in spatial relations between baboon
mother and infant were made by the infant more often than by its
mother. During the first few months of life, however, infants increased
the distance between the two and mothers decreased it. At this time
infants were about four times as likely to leave their mothers as their
mothers were to leave them, with considerable individual variability.
Later it was mothers who became leavers, infants approachers, as has
also been reported in laboratory investigations of a number of primate
species (e.g., Hinde et al. 1964, Hinde and Spencer-Booth 1967,
1971, Jensen et al. 1967, Kaufman 1974, Rheingold and Eckerman
1970, Rosenblum 1974, Rowell et al. 1968) and by Nash (1978) for a
few older baboon infants at Gombe and by Berman (1978) for free-
ranging rhesus macaques of the Cayo Santiago colony. In the present

study mothers and their older infants were equally likely to leave each other, but reunion was nine times as likely to be due to the infants as to their mothers.

To facilitate comparisons I have excluded from the "increases" and "decreases" those transitions that were between contact and arm's reach to obtain in the "approaches" (%A) and "leaves" (%L) of Hinde and his colleagues (Fig. 61). The basic graph remains unchanged from that for decreases and increases, but a few differences bear comment. The main difference is that during the first month of infant life most of an infant's increases in distance from its mother take it no farther than her arm's reach. When these are excluded, the infant's contribution to leaving drops from over 80 percent to just over 60 percent. Moreover, the young infant's contribution, in general, to both approaches and leaves is about 10 percent less than when transitions between contact and arm's reach were included. Results past the fourth month of life are indistinguishable from those noted previously, with infants making 90 percent of the leaves from month four on and about 55 percent of the approaches from month five or six on.

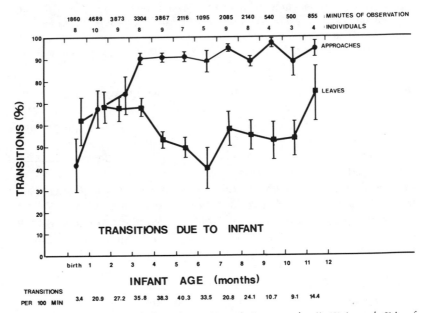

Fig. 61. *Percentage of the "leaves" and "approaches" (%A and %L of Hinde and coworkers) by mother or infant that were performed by the infant at various ages. The average rate (acts per 100 minutes) at each age is indicated below the horizontal axis.*

Infants become leavers more than approachers in the second or third month for the baboons in Amboseli, the provisioned free-ranging rhesus of Cayo Santiago, and the caged rhesus in the Madingley colony (see comparison in Berman 1978), a rather amazing amount of similarity in results between species and between settings. The rhesus in both settings reach a constant level of %A–%L at about 12–15 weeks of age, and about a month later for the Amboseli baboons. Also, the asymptotic level is approximately 35 percent in Amboseli, 40 percent on Cayo, and only 10–20 percent in Madingley. That is, infant contributions to spatial proximity stabilized later in wild baboons than in captive or free-ranging rhesus and at the stable level there was much less difference between an infant's contribution to leaving and its contribution to approaching in the captive rhesus than there was in either the free-ranging rhesus or the wild baboons. From the published data it is not possible to analyze further the source of this difference.

Of particular importance is the contingency found between infant contact and various maternal activities. I suggest that this contingency is at least partially imposed upon the infant; that is, the infant must adjust its behavior to that of its mother rather than vice versa (the main exception being the mother's adjustment of feeding postures described in Chapter 5). The young infant learns not just that it can be out of contact sometimes but that it can be out of contact at particular times that are determined by its mother's activity. This is probably one of the earliest forms of socialization for an infant baboon and may provide the social sensitivity on which later social integration depends. One of the first areas of infant trauma seems to involve the reversal in these contact contingencies. Older infants must take advantage of their mothers' rest time for any contact that they want rather than using that time for play and exploration. Thus at one level maternal rest time remains the same: it is the time to do the "unusual". However, the unusual is lack of contact for young infants and contact for older ones. Thus at the level of immediate behavior it is a reversal. At four to six months of age the contingency reversal seems to precipitate some of the first infant "weaning," or distress, behavior. In the next chapter I shall consider various aspect of independence and weaning in more detail.

9 / Weaning and Infant Independence

PRIMATE MOTHERS PROVIDE their young with a number of important kinds of care. Infant development and independence involve the infant's gradual attainment of self-sufficiency in each of these areas. In the scientific literature, as in common English usage, the term "weaning" is sometimes used to describe this overall process of increasing independence. However, the term also retains its narrow and original reference to only nutritional self-sufficiency from mother's milk and to self-sufficiency that is imposed rather than voluntary. I think it is important, if a bit more linguistically awkward, to consider separately the various areas of care and to leave open to theoretical consideration and empirical study the question of the behavioral source and imposition of independence in each area.

Initially, baboon mothers provide all their infants' nutrition through nursing. They provide all transport through primarily passive carrying of the clinging infant. Protection from disease and ectoparasites is accomplished through grooming and perhaps through restriction of early exploration and of social contacts. Protection from predators is provided at first by carrying the infant away from danger as the mother herself detects predators, responds to detection by other group members, and either flees or joins in group action against predators.

An infant's increasing size and independence affect its requirements in each of these various areas. For the infant, its mother, and for other group members, the cost and the individual's ability to provide care for the infant will be a function of the type of care and the characteristics of the potential provider: there are alternative ways of meeting requirements, alternatives that involve different individuals and different behaviors. Below, I shall outline these parameters for baboons, providing what I hope is the first step toward a complete specification

168

of costs, benefits, and options that will lead to similar research on this and other species.

Transportation

The available evidence in the literature suggests that the energetic cost of carrying an infant is essentially proportional to the additional weight involved (Givoni and Goldman 1971, Soule et al. 1978). One might at first assume that it is advantageous to a mother to stop providing her infant's transportation as soon as the infant can keep up with the group on its own, and in fact, as noted in the previous chapter, infants spent considerable time out of contact even during travel at ages at which they were still primarily dependent on their mothers for food. However, a small animal in general expends more energy than a large one in covering a given distance in a given amount of time (Taylor et al. 1970). Therefore, if a mother is the sole source of nutrition for her infant, it may cost her more if the infant becomes semi-independent and provides some of its own transport through locomotion than it would cost her to carry it. We cannot assume that apparent independence is advantageous to a mother and costly to her infant. The true situation may be the opposite of the apparent one.

Carrying is an area of care in which other group members could readily substitute for mothers, a behavior that occurs to varying degrees in other primate species including humans (see Konner 1976, Kummer 1968, Redican 1976). Provision of transport by other group members is relatively rare in savannah baboons, although, as noted elsewhere, in the most extreme case of maternal rejection during this study (Gin's rejection of Grendel), the infant spent considerable time riding on adult male High Tail, and after High Tail's death unsuccessfully sought rides from other group members until many months later he began to be carried by an infertile adult female, Lulu. Among those whom one would most expect to provide care are siblings, but they are quite likely to be immatures in this slowly maturing species. A six-month-old infant would impose an appreciable additional energetic cost on a half-grown juvenile sibling three years old. Four- to five-year-old siblings are probably large enough to comfortably provide some transportation for infants. The only members of this age class during 1975–76 were immigrant male Nog and males Toto and Dogo, neither of whom carried their siblings, Eno and Pedro. Almost all adult female relatives will be either pregnant or supporting their own infants. This leaves subadult and adult males as the most likely candidates for such care, and the adults do provide some. Perhaps the adults would provide more infant care in a more structured mating system than that

Fig. 62.   *Grendel trying to obtain a ride from two-and-a-half-year-old Cete.*

found in these baboons, that is, one in which individuals are more (predictably) closely related to any particular infant in the group (see discussion in Kleiman 1977, Redican 1976). It is interesting in this context that adult male hamadryas baboons, who maintain long-term polygynous bonds, establish their first harem relationships by providing much of the transportation for young juvenile females (Kummer 1968, Abegglen 1976). Immature hamadryas probably require carrying at an older age because the groups travel longer distances in a day than do Amboseli baboons. It seems likely that for any primate species the amount of observed kin involvement in infant care at any particular time will depend on the age, sex, and kin structure of the social group at that time and on the other demands placed on potential helpers.

Nutrition

As indicated previously (Chapter 5), caloric requirements to maintain weight are probably proportional to weight to the three-quarters power. However, a growing infant also requires additional calories for the production of new tissue. Additionally, to the extent that an infant is more active when it is not in contact with its mother, to that extent caloric requirements will be greater in order to maintain the increased activity level. Thus from the standpoint of a mother who is providing all her infant's nutritional requirements, it is preferable for the infant to remain totally dependent, in passive contact, rather than to be semi-independent, running about.

Would a mother gain as much advantage for her infant, at less cost to herself, if she allowed it to be active and out of contact and to obtain some of its food independently? Only if the assimilable calories the infant so obtained were greater than the infant's increased caloric expenditure in obtaining those calories and in running rather than being carried (see "Transportation," above). Although this is obvious, the practical implications may not be. As I discussed previously, many foods just are not accessible to harvest or to assimilation at some times of the year or before infants are at certain stages of physical development. Thus an infant's physical development and the availability of appropriate weaning foods determine whether alternatives exist, alternatives that would be advantageous to the mother as well as to the infant. In the short run, it could sometimes be more costly to the mother, just in energy considerations, for an infant to try to obtain some of its own food than for the mother to provide it all through lactation. It is possible, of course, that there are some essential nutrients that the infant requires past a certain age that are not obtainable from milk and that are available in weaning foods. These and similar constraints favoring either milk or other foods would have to be taken into account in complete analysis of the economics of weaning.

Two other alternatives are used by some other animals to provide food for young that are unable to provide it for themselves—harvesting and/or partially processing foods for the infants. These systems have the advantages that other group members can share with the mother in the provision of food for the young and that perhaps less energy is lost than if the food is converted to milk. Aside from humans, most vertebrates that use these feeding systems have two common features—first, the use of a den, nest, or other predictable location where the young are kept and to which the feeders return with food, and second, the exclusive use of animal rather than vegetable foods. Perhaps vegetable foods are not as suitable for animals to harvest for other individuals unless the food can be cooked, as it is by humans. Certainly, for a baboon group, which forages over a long day journey and feeds throughout the day, such a feeding scheme would be more difficult. Perhaps these are some of the reasons that no such feeding system has developed in baboons, but the answer is not totally satisfying, and this seems to be an area that warrants further study. The only way that group members provide food for a baboon infant, other than its mother via lactation, is through tolerance of infants at a food source, the result of which is that infants obtain scraps of foods that they themselves

Fig. 63.   *Grendel reaches for his mother's nipple as Gin feeds. Unlike a young infant, one undergoing weaning often first reaches for the nipple by hand, moving its head to a nursing position only if not rebuffed at the first stage.*

could not obtain by their own effort (Fig. 64). This more subtle, passive form of food sharing may be a very important one for a number of animals and is one that has received less attention than more dramatic forms just because of its subtlety. This is a common problem in behavioral research—dramatic behavior receives more attention.

Disease and Ectoparasites

It would at first seem that an infant's need for grooming would increase more slowly than its nutritional and transport requirements because surface area increases more slowly than does weight or volume. However, an infant that is semi-independent, that is close to the ground and is running and playing in shrubs and grasses, probably has a higher density of ectoparasites than one that rides on its mother. The time required to free a semi-independent infant of ectoparasites is therefore probably greater, proportional to the infant's weight, than the time required for a totally dependent infant. Again, we have the suggestion that there are certain advantages to a mother in having her infant clinging all the time rather than being semi-independent. The infant itself cannot readily do its own grooming: it is not physically able to perform complete grooming manipulations until at least eight

Fig. 64. *Five-month-old Safi eats a scrap of a corm her mother has dug.*

Fig. 65. *Adult male Peter threatening an infant that moved too close while trying to obtain part of a lizard Peter had caught.*

months of age, and there are many parts of a baboon's own body that even an adult cannot reach to groom. Grooming appears to be a function that could easily be assumed by animals other than the mother. Others can and do groom infants, infants of laissez-faire mothers in particular, but most of a young infant's grooming is done by its mother, often at the same time that the infant is suckling. Perhaps this cements the bond that leads later to the considerable reciprocity of grooming relationships that is seen among juveniles and their mothers. That is, mothers may provide most of their infants' grooming in part to obtain even more grooming in return when their infants mature.

To the extent that exposure to disease is a function of environmental and social contacts, it is also a function of time spent away from the mother, but probably is otherwise essentially independent of infant age or size. Mothers cannot effectively control their infants' exposure when they are out of contact and can only partially do so when they are in contact. While they are in contact, mothers provide some regulation of social interaction, as described in previous chapters. However, I have not seen mothers limit, for example, the nonfood items that their infants mouth from the ground, especially in their first weeks. Note that it may be preferable for baboons to be exposed to certain diseases as infants rather than as adults, as with the so-called childhood diseases in humans. Again, it may prove difficult to compare properly even relative costs or benefits of behavioral alternatives.

Protection from Predation

With protection from predation as with most of the areas of care that I discuss, it is harder, not easier, for a mother to provide care if her infant is not in contact, no matter what the size of the infant. That is, the same benefit to the infant is more costly for the mother to provide if her infant is not in contact. A mother must first locate and make contact with her infant before carrying it in flight. Additionally, if the two are separated, it would seem that infants can more readily keep track of and make contact with their mothers than vice versa, because mothers are larger, generally slower-moving individuals whose activities and location are probably more predictable. Thus maintaining information about the other's location is probably done at little cost by the infant but at high cost by its mother. This theoretical speculation was supported by several kinds of observations. I already described the sudden appearance of older infants when their mothers began to rest. Additionally, in alarm situations, I commonly saw mothers of semi-independent infants stop, look about, appearing unsure of their infants' location, and then wait as their infants dashed out from a bush or tree (I had not been able to locate the infant) and approached. The mother sometimes met the infant as it came close and in every case she embraced the infant to contact, and then dashed away. As infants stray farther and farther from their mothers, even this cooperative form of flight probably becomes inefficient for both. We do not know at what age an infant can run, for at least short distances, faster than its mother, or at least faster than its mother can run while carrying the infant. Also, by seven or eight months of age infants can climb even fairly tall fever trees. Ten- to twelve-month-olds were often seen to run alongside their mothers in flight when the group fled from a predator or local Maasai tribesmen. It was also at these times of emergency that particular associated adult males were seen to carry infants and young juveniles (see Chapter 6 and Appendix 3).

"Weaning": Maternal Punishment and Infant Distress

In the present study detectable aggression (usually biting, pushing, grabbing, or hitting) directed by a mother toward her infant was very rare, for most mothers and most infant ages, less than one act per 100 minutes. Only Gin and Spot aggressed against their infants during month one; Gin in month two; and Gin, Spot, and Preg during month three. Yet by month five all mothers had bitten, pushed, grabbed, or hit their infants. Maternal aggressive or punishing behavior was commonly followed by gestures or vocalizations of distress by the infant, particularly grimaces and the vocalizations "eee," "ikk," and "coo" (Appendix 4). In extreme cases, infant distress took the form of a tantrum (Fig. 41), very much resembling those thrown by two-year-olds in our society. An infant baboon throws itself to the ground in a "fear paralysis" (Appendix 4), screeching ("eee") and cackling ("ikk"), while waving its tail about and watching its mother, who usually ignores the infant. The infant may periodically run after her as she forages and the infant also sometimes interjects long "coo" vocalizations. This dramatic display sometimes lasts at high intensity for five to ten minutes, recurs sporadically throughout the day, and leaves infants thoroughly exhausted afterward. Although these extensive tantrums as well as their briefer or milder counterparts do sometimes follow aggression or punishment by the mother, they occur five to ten times as often as do such maternal acts. More commonly, a mother causes such infant distress merely by just getting up and walking away when her infant tries either to make contact or to suckle. Or she shifts position slightly, making the nipple inaccessible. Or, if the infant climbs on her back while she is walking, she suddenly sits down, thereby knocking the infant off. It is very hard to distinguish objectively these subtle acts from a mother's normal activities except from the subsequent infant behavior, a criterion that would become quite circular. However, this behavior by mothers does seem to be quite common during about months four to six, and I think that using cine film or videotapes from that period played back at slow speed might help obtain a better understanding of the communication between the two.

What seemed to stimulate punishing or aggressive acts by mothers toward their infants? During the first few months, a mother gently withdrew her arm if her infant tried to hold her food or arm as she fed. Later she pushed the interfering infant away. Finally, mothers roughly shoved away older infants, who were threatened away if they even tried to feed within the mother's arm's reach. Occasionally (frequently for Grendel) infants were specifically rebuffed from nipple contact when they were four to six months of age, but most of these rejections seemed mild and situational rather than general—mothers often allowed long nursing sessions shortly after a brief rebuff. As pointed out in Chapter 8, much so-called weaning behavior actually serves to condition the infant to the "proper" time for contact and nursing. Consistent with this are limited data on the time infants spent on the nipple, which indicate that "nipple time" dropped rapidly over the first four months without rejection, decreased no more rapidly in the next few months. Also, infants and young juveniles continued to spend some daytime on the nipple until part way through their mother's next pregnancy, usually in the offspring's second year, and perhaps considerable time on the nipple at night—yearlings still slept clutched in the mother's ventrum at that age. No data are available on the amount of milk obtained by older infants, but it seems likely that at least some milk is produced as long as some sucking occurs and the mother does not become pregnant. Recall, however, that I could only occasionally identify either sucking or swallowing. Usually, I could only determine that the nipple was held in the mouth in the usual nursing position. In this study, occasional overt nipple withdrawal and punishment occurred even at resting times and sometimes after the infant had been on the nipple for a considerable time. At other times infant distress seemed to be due not to punishment but to absence of milk; the infant repeatedly switched nipples and often pulled one or both nipples by hand with no interference by the mother. This pulling may stimulate lactation.

Maternal punishment also occurred when five- to eight-month-old infants repeatedly hopped on and off mothers, and mothers sometimes shook infants off when they were riding dorsally. At the extremes were Gin, who hardly tolerated any riding even when Grendel was

only four months old, and Scar, who did not resist Summer's very frequent hopping on and off Scar's back whenever Scar took a step or,
alternatively, paused for a moment. Only during months seven and
eight did Scar begin to lower her hips or shake them slightly until Summer jumped off.

A situation that often stimulated loud, persistent infant distress
occurred during descent from the sleeping trees in the morning. During month seven most mothers descended the sleeping trees without
their infants, leaving them behind. At first infants protested violently
and mothers sometimes returned to sit halfway up the tree, waiting
until the infant descended that far, and then carried it the rest of the
way. Occasionally an infant fell in its attempts to descend. At other
times infants obtained rides on their adult male associate. Mothers
often sat behind the group, facing and watching the sleeping tree until
the screaming infant descended, at which time the infant usually ran to
its mother, rode, and nursed. This forced independence period lasted
only a month or less; eight- or nine-month-olds almost invariably descended alone in the morning without signs of distress.

Summary of Weaning and Gradual Independence

In the previous sections I have tried to indicate the variables affecting infants' requirements for care in several areas in which all care
is initially provided entirely by their mothers. I have also tried to indicate possible alternatives to maternal care that are available at various
ages and the effects of growth and of spatial independence on both the
infant's requirements and the mother's ability to satisfy those requirements. Our knowledge of the time course of each of these is clearly
imperfect. My main hope is that outlining both the information necessary to understand the development of parent-offspring relationships
and our current state of ignorance will stimulate the gathering of the
necessary field and experimental data.

Despite the gaps in current knowledge, we can draw several tentative conclusions at this stage. Most of an infant's requirements for
care increase proportionately with the infant's increase in weight. In
addition, both the requirements and the cost to the mother of meeting
the requirements also increase as a function of the time an infant
spends actively out of contact with its mother. Other conspecifics contribute, but apparently little, to providing the necessary care. If we ignore increased requirements due to semi-independence and consider
only the effects of weight gain, even at slow rates of infant growth, that
is, 5 grams per day, provision of just an infant's nutritional requirements and transportation would appreciably tax a mother's time bud-

get by the time her infant was six months old (Chapter 5). Maternal punishment and rejection during the infants' fourth to sixth month results at first in dramatic tantrums, but infants soon seem to learn to restructure their schedules so that they obtain care without interfering with their mothers' major maintenance activities.

One interesting speculation is that low-ranking mothers, by being restrictive and as a consequence spending more time in contact with their infants, provide less costly care for their infants during the months of total nutritional dependence. If so, they might, thereby, counteract the effects of being supplanted from food resources more frequently than high-ranking females are, which would otherwise necessitate their spending more time feeding. This would account for my inability to detect any relationship between a mother's dominance rank and the time she spent feeding.

Evolutionary biologists propose that these energetic, attentional, and time budget costs and protection from sources of mortality are translated into costs of and benefits to survival and future reproduction. On this basis, and assuming a measure of heritability of the traits involved, they have considered the potential for the evolution of such traits and have in addition proposed evolution through natural selection as a potential explanation for the development of observed social behaviors. Although the basic ideas were formally proposed at least as early as 1930 by Fisher, it is due primarily to expositions by Alexander (1974), Trivers (1972, 1974), West Eberhard (1975), and Wilson (1975), based on the work of Hamilton (1964), that these ideas increasingly have been applied as explanatory devices to the behavior of primates, including humans. In the next section I shall review the major features of both the more heuristic and the formal population genetic models, particularly as they apply to parent-offspring and sibling interactions, and the empirical evidence from primates that bears on these models.

Evolutionary Models of Parental Investment and Parent-Offspring Conflict

Robert Trivers, in a particularly stimulating and influential pair of papers (1972, 1974) developed a line of arguments based on pedigrees to produce predictions regarding the nature and time course of parental behaviors as a function of costs to the parents' future reproductive success and of benefits to survival of the current infant. Trivers' basic argument was that because parents are in general equally related to all their offspring but any particular (current) offspring is more closely related to itself than to its siblings, natural selection will favor offspring who attempt to obtain more parental investment (behavior that increases the survival of the current offspring while reducing a parent's ability to invest in future offspring) than their parents are selected to provide (Trivers 1974). Trivers further suggested that this genetic "conflict of interest" is resolved by behavioral conflict between parent and offspring. He pointed to the existence of behavioral conflict between parents and offspring, particularly in humans and baboons (Trivers 1974), as support for the idea that genetic conflict of interest has produced parent-offspring behavioral conflict. Clearly, there are several different questions that are relevant to this sequence of arguments.

Could a gene spread that caused an offspring to obtain more parental investment than its parent was selected to give? If so, under what conditions? Do the existing conditions for any particular species allow for the evolution of such offspring behavior in that species?

Is behavioral conflict the inevitable outcome of genetic conflict of interest? Are there reasonable behavioral options that provide more favorable cost-benefit ratios and/or require more probable genetic changes than does behavioral conflict?

Are there explanatory mechanisms for parent-offspring behavioral conflict that provide viable alternatives to the evolution through natural selection that is due to genetic conflict of interest? I shall consider each of these in turn, particularly as they apply to primates.

Several authors (e.g., Charlesworth 1978, Charnov 1977, Levitt 1975, Wade 1978) have recently constructed population genetics models of altruism, particularly sibling altruism. Each has assumed some aspects of large, randomly mating populations with large sibships. These assumptions apparently are approximately satisfied by some insect populations.

In 1978 three papers (McNair and Parker 1978, Parker and McNair 1978, Stamps et al. 1978) appeared in which population genetic models of parent-offspring conflict were presented. These followed on a note published the previous year (Blick 1977), the first attempt, to my knowledge, to model formally Trivers' (1974) suggestion that parent-offspring conflict will evolve and Alexander's (1974) rebuttal that this is not possible (see also Dawkins' 1976 rebuttal to Alexander). Alexander argued that such genes will be selected against and that selection for parental manipulation of offspring will evolve. As with the related genetic models of sibling altruism, each of these recent population genetic models of parent-offspring conflict includes slightly different sets of assumptions and modes of analysis, which make it difficult to interpret results that differ. However, each model does demonstrate some conditions under which parent-offspring genetic conflict of interest could occur and that behavioral conflict might evolve through natural selection.

However, a possible problem arises when we attempt to apply either the existing sibling or the parent-offspring models to primate evolution. Primate sibships, and those of a number of other large mammals, differ considerably from those assumed in most models. First, sibships in these species are very small, probably less than ten for most species. Second, the individuals constituting a sibship are not members of a single litter or even two or three litters, but rather are born singly, spaced over fairly long time spans. This results in lower levels of overlap between siblings at any life stage such as the period of infantile dependency, and less overlap throughout their lifetimes, producing an even smaller "effective sibship size" (in the same sense as "effective population size" is used in population genetics). Further, to a considerable extent maternal siblings will be half-siblings in many species. Paternal sibships also may be discrete, potentially interacting units, for example, age cohorts, under some mating and social conditions (J. Altmann 1979), but they will characteristically be different from maternal sibships, offering opportunities for sibling interactions different from those offered by maternal sibships.

Sibship size and the number of offspring produced by any mating combination seem to affect the assumptions of or appear in the formal models in several places. It is not obvious to me how robust the models are under the conditions that are produced by most primate systems of mating, reproduction, and rates of maturation. I have not seen a treatment of this in the literature, although Charnov (1977, corrected in Johnson 1979) and Stamps et al. (1978) consider several special cases in which certain aspects of small sibships are accommodated in particular formulations. I hope that these points will be addressed and clarified so that appropriate applications of the models to these systems can be pursued.

An additional question arises when we consider estimation of the parameters of the models. Crucial both to the earlier formulations and to their more formal population genetic counterparts are measures of costs of an act, usually to the actor (e.g., parent), and benefits of that act, usually to the recipient (e.g., infant), both costs and benefits to be measured in units of survival and reproductive success or reproductive value. What are the costs and benefits of maternal care?

If a current infant garnered more care than its mother "was selected to provide," there are two major ways in which a primate infant might reduce its mother's future reproductive success: either through increasing the probability of her death or through delaying the conception of her subsequent offspring.

Greater mortality risk might arise from several sources. If a female increases the attention she devotes to locating food and to keeping track of her infant, I suggest that she has less attention available (see Kahneman 1973) for predator detection and is more dependent on the alarms of the other group members. Increased infant care might also render a mother more susceptible to predation if she were less able to keep up with the group and therefore stayed at the rear, as we have commonly observed, and were therefore more susceptible to predation (Hamilton 1971), particularly so if she could not flee from a predator as fast because she had to retrieve the infant (if it was out of contact) and if she ran more slowly due to the greater weight, or just due to poor physical condition.

The data from Chapter 6 further suggest that greater mother-infant contact leads to greater contact between the mother and other group members, which in turn, particularly for low-ranking females, leads to increased stress and greater energy demands, all of which would render a female more susceptible to diseases caused by bacteria and by viruses. It also is possible that mothers are in a negative energy balance (Chapter 5), and may accumulate particular nutritional deficiencies as well, perhaps resulting directly in starvation or, more likely, nutritional diseases.

Delay of future reproduction may be accomplished through the direct hormonal effects of sucking or through nutritional and other stress, perhaps including weight loss. This is the one area in which evidence is most available for at least the broad outlines of apparent reproductive costs (see J. Altmann et al. 1978, Jain et al. 1970, Lee 1978, Saxena 1977). Yet the information is still inadequate for determining the effects of differences in amounts of maternal care of infants at various developmental stages on the timing of subsequent reproduction in wild primates.

The question then arises, could mothers do more? The form of maternal investment most often considered is nutritional, through lactation, and the most obvious form of parent-offspring conflict is thought to be nutritional weaning. Initially, a mother provides all its infant's energetic needs, primarily through lactation. Is she being selfish not to continue to do so? Consider a baboon mother who ordinarily spends over 55 percent of her time feeding, approximately 23 percent walking, 20 percent resting or engaged in grooming interactions when her infant is about five months old. At this stage, her infant spends 30 percent of its time in contact with her. This is the age at which pronounced dramatic tantrums occur. It is six months before most mothers resume cycling and nine to twelve months before most become pregnant again. It is most unlikely that the mother of even a six- to nine-month-old could provide enough care for survival of the infant and also support the strains of a new pregnancy. Let us assume that a mother "decides" to nurse her infant more. In order to do so she will have to spend more time feeding. Yet the sum of the mother's resting, grooming, and walking time will be decreased if the time spent in the one remaining activity, feeding, is increased. Furthermore, walking time is primarily determined by the movements of the whole social group during its day route rather than being subject to appreciable individual variability. Thus it is the mother's resting and grooming that will suffer. To the extent that these activities usually are beneficial, there will probably be a loss to both mother and infant.

Deleterious changes in allocation of time, energy, or attention, if they have life history consequences at all, are likely to have immediate ones in terms of reduced chances of survival for the mother or her current infant. In the absence of evidence to the contrary, it seems more reasonable to assume that the effects on biological fitness of an act are usually a nonincreasing function of the time since the act occurred. Costs to the mother are therefore likely to be either immediate reduction in the chance of surviving or diminished ability to provide care for the current infant, for example, reduced feeding efficiency, reduced attentiveness to predators, or inability to keep up with the group. In this context it is important to remember that the current infant's survival is entirely contingent on its mother's survival: in all cases of maternal death in baboons, the infant (i.e. under a year of age) has also disappeared. In Trivers' terminology I am proposing that what he called "self-inflicted" costs to the infant are greater and play a more important role in parent-offspring relations than may have been assumed. In terms of the population genetics models, I am suggesting that the benefit of maternal care to the infant is not a linear function of cost to the mother but rather approximates a linear function of the log of cost and reaches an asymptote or upper bound.

These immediately detrimental effects of any attempt by an infant to demand more of its mother, combined with an infant's increasing abilities to care for itself, abilities whose development is crucial to the infant's eventual independence, suggest that parent-offspring genetic conflict of interest may arise infrequently as a relevant variable in many real-life situations.

Trivers further predicted that maternal care should be a direct function of maternal age. This argument was based on reproductive value being a monotonic, nonincreasing function of age for adult females and on the assumption that the relative "worth" of each of a parent's offspring is determined only by the offspring's genetic relatedness to the parent. However, additional factors should be considered in evaluating the worth to a mother of her various offspring.

1. Survival of early offspring versus later ones represents a reduction in generation time and therefore, if increased maternal care improves infant survival, greater maternal care for early infants would be favored by natural selection unless the population is a decreasing one (Cole 1954, Lewontin 1965, Mertz 1971). Note, too, that a surviving six-month-old, having survived a life stage of high mortality, has a much higher reproductive value than a future unborn offspring.

2. The potential of early offspring for "help at the nest," sibling altruism, and more reciprocal altruism with the parent due to a greater overlap of lifetimes, would also lead to a prediction of greater maternal care for early versus late offspring.

3. Mortality rates are a nondecreasing function of age in mammalian adults (Caughley 1966). Thus a later offspring is more likely to be orphaned, so it is more important for its survival for it to become self-sufficient at a younger age. To the extent that care beyond a certain level retards development of self-sufficiency, such care would be disadvantageous to the infant.

4. Because of decline in metabolism and other effects of aging, the immediate costs to an older female of providing a given amount of infant care are likely to be greater than those for a younger female, offsetting any decreased long-term costs as a result of lower reproductive value.

These factors can be subsumed in the cost-benefit ratios of the various models, and in that sense they offer no problem to the models as they are generally established. However, they do affect the forms that the functions have been assumed to take. These factors also affect benefit to the current offspring or cost to the parent (see Emlen 1970, West Eberhard 1975). For example, these effects are such that they produce predictions of more care for early offspring, that is, decreasing maternal care with increasing maternal age. Effects 1 and 2 are particularly compelling. Although effects 3 and 4 are more debatable than the other two, they are still plausible. In any case, we are left with the need to state explicitly a larger set of assumptions that have often been implicit, to measure the costs and benefits and, perhaps, to develop more appropriate genetic models before we shall be able to make evolutionary predictions relating, for example, parental care and parental investment to parental age.

Will conflict of interest always lead to behavioral conflict? I have suggested that more often than supposed there will be no conflict of interest between a parent and its current offspring regarding the amount of care the parent should provide. It is still important to ask, what will happen when conditions of conflict of interest arise? Is behavioral conflict the inevitable resolution of genetic conflict of interest? The answer depends on the net cost to each participant in the conflict and on the net costs of available alternative behavioral strategies.

Behavioral conflict itself probably incurs a cost. Baboon infants that throw long tantrums lose time from feeding and other activities during that time, and they appear to be thoroughly exhausted when they finally stop. Moreover, it has been suggested (e.g., by Trivers 1974) that loud vocalizing by baboons renders these individuals more susceptible to predators. Alternatively, an infant might obtain a greater net gain (benefit) by waiting to obtain more milk from its mother when she is resting than by fighting (often unsuccessfully) to be allowed on the nipple when she is feeding. Providing additional care at rest time will cost the mother less, and so she will be more willing to provide it. Such less costly care is lower investment in Trivers' terms than care provided when the mother is feeding, but it may provide the infant with as great, or greater, benefit than care it forced its mother to provide while she was feeding or walking, both because the infant would not suffer the cost of the behavioral conflict itself and because the infant would suffer fewer self-inflicted costs resulting from its effects on its mother's survival.

Essentially, I am suggesting that an infant often can gain a particular amount of benefit in several different ways, which will vary in their cost to its mother not in just one way at a fixed single cost, and that these alternatives, along with the costs of behavioral conflict, will determine the nature of the behavioral resolution of potential conflict situations. The restructuring of contact time, described in Chapter 8, appears to be a behavioral resolution through change rather than conflict. Such resolutions may often be the optimal ones.

But having chosen nonconflict strategy in some situation, wouldn't an infant still try to obtain even more care? That is, wouldn't the older infant that learned to seek out nursing when its mother was being groomed or the human toddler who learned to wait to be held until its mother finished a chore and sat down to relax, still try to obtain somewhat more than its mother "wanted" to provide? Perhaps, but not necessarily. Again, costs and benefits of alternative strategies must be evaluated. The additional investment just obtained places the current investment level at a different, higher value on the cost-benefit curve than previously, and if the curve is horizontally asymptotic, discontinuous, or has other features already indicated as likely in real-life situations, then the self-inflicted costs may be greater than the potential gain that was being disputed.

In sum, conflicts of interest do not invariably lead to conflicts at the behavioral level. Cooperation and compromise are likely alternative behavioral resolutions of conflicts of interest. The relative use of different behavioral options will be a function both of the values of the alternatives and of the animals' ability to make the relevant evaluations of strategy. In a socially living animal, cooperation and compromise in parent-offspring interactions may form a useful developmental base for group organization and cohesion among adults and serve as a model for resolution of potential conflict situations among adults. In contrast, resolution by conflict may be more generally appropriate to more solitary and territorial species.

Chase (1980) has pointed out that most existing models and theories dealing with the evolution of behavior have dealt with existence statements—can a behavior evolve or not—rather than with apportioning time or making choices among various behavioral alternatives. Maynard-Smith's analysis (e.g., Maynard-Smith and Parker 1976) of evolutionary stable strategies, based on a game theoretic approach, is one way of considering several viable alternatives, and Chase makes use of more general economic models to propose others. Recent interest in incorporation of time as a valuable and limited resource in models of human economic decisions (e.g., Becker 1965) may prove useful in biology. The future development of models of alternative options should provide another fruitful way of looking at the evolution of social behavior in general and parent-offspring relations in particular.

Behavioral conflict certainly may arise as the resolution to genetic conflict of interest. However, caution will be needed in predicting the situations in which it is to be expected, and exclusive focus on this form of resolution would certainly be misleading. Is similar caution needed in attributing observed conflict to genetic conflict of interest?

Behavioral conflict between parents and offspring of several primates, including humans, is an observed phenomenon that is sometimes quite dramatic and attention-attracting. Is it always a result of genetic conflict of interest? The classic tantrums of two-year-olds in our culture and the tantrums of five-month-old baboon infants (Fig. 41) provided the prime examples in Trivers' (1974) work. The theoretical discussions have focused on conflict in which the issue between parents and offspring seems to be the amount of care the parent will provide, with the infant seeking "more," the parent "less." However, in baboons as in a number of other species, it is the infant who initially seeks independence and the parent who restricts it (see e.g., J. Altmann 1978, Hinde and Atkinson 1970, Rheingold and Eckerman 1970); the result is often struggles. Rowell (Rowell et al. 1968) first described restrictive baboon mothering in her caged colony of baboons and speculated that it might be a product of captivity, but the present study has shown it to be a more important and general phenomenon, as it apparently is in macaques (see the review in Berman 1978). Even conflict at older ages does not always fit easily into a parental-investment model, nor can we reject a priori alternative explanations. In any species for which there is a long period of dependency and in which we acknowledge the large role of learning during the lifetime, it seems important to consider the potential value of learning in parent-offspring relations.

This flexibility allows for developmental fine-tuning based on initial and changing states of both mother and infant, which in turn are dependent on a variety of ecological, maturational, and social factors, as shown in previous chapters. Even if a mother's and infant's "interests" overlapped entirely, owing to different environmental histories, the two are, as a dyad, a new phenotypic combination just as they are new genetic combinations as diploid individuals. A mother will need to alter her care depending on the season of her infant's birth, the prevailing conditions when her infant is at various stages of development, characteristics of the infant such as health, and the particular demographic and social milieu. If these factors were constant over a female's lifetime and over a number of generations, then perhaps a more "fixed program" (Mayr 1974) would have evolved as a more efficient system (see Bateson 1963, Levins 1968). If conditions were constant over a lifetime, then perhaps some sensitive period during juvenile years or the time a female spends caring for her first infant would completely determine a female's investment pattern thereafter. Variability within a lifetime, however, will require repeated adjustments. A mother is to its infant and an infant is to its mother one of many only partially predictable environmenal variables, not a predetermined constant. Additionally, their fates are inextricably bound, probably more so than those of any other pair of individuals or environment-individual pair, yet it is unlikely that they could have perfect knowledge of each other either initially or during development. Feedback mechanisms are necessary within each and between the two to monitor their own and each other's often rapidly changing states, particularly those due to maturation. Thus in addition to the possibility that behavioral conflict may sometimes be a consequence of genetic conflict of interest, it also may have evolved without genetic conflict of interest as a result of the infant's immaturity at birth, its long period of dependency, and the changing conditions under which parenting and development take place over a lifetime.

10 / Conclusions and Speculations

Maternal Care and the Infant's First Few Months of Life

A BABOON INFANT is born into its mother's world, and immediately that infant is affected by those things that affect its mother. The infant also intensifies the very factors that influence her, as it draws the attention of other group members and increases the time and attention she must devote to maintenance activities. The adult males with whom she previously associated usually are the ones with whom the infant associates. The mother's adult female and juvenile associates also tend to be the ones who associate with the mother-infant pair, but in addition there are others of these classes who are just interested in all infants and who interact with mothers with whom they have not had previous association. The attraction of others to the infant results in low-ranking mothers being put in fearful, stressful situations much more often than otherwise because they can no longer keep a distance from their higher-ranking peers. Because infants are often pulled and even kidnapped by others, these low-ranking mothers are much more restrictive of their infants' movements, and the result is that their infants remain dependent on the mothers several months longer than do their peers.

Can we identify the characteristics of "good" mothering and "bad"? If we use a biological criterion of survival to maturity and first examine infant survival, data from this small sample of mothers suggest that restrictive mothering is perhaps better mothering. However, we must be cautious not only because of the sample size but because of two other factors as well. First, it seems likely to me that restrictive mothering may result in higher survival during the early months because it probably reduces the chances of falls, disease, kidnapping, and predation. However, I have shown that infants with restrictive

187

mothers develop independence more slowly. Thus it also seems likely that offspring of restrictive mothers will be less able to survive their mother's death at an age that offspring of rejecting mothers could; for example, the former infants might be able to survive alone only after a year and a half, the latter by one year of age. The particular balance of mortality pressures between those producing early infant mortality and those causing maternal death might vary from year to year or because of environmental differences. Restrictiveness is a strategy that achieves a short-term gain at the cost of a long-term loss.

If this basic analysis is valid, one would predict that mothers will tend to become more laissez-faire as they age because their later offspring are more likely to be orphaned, and early independence would therefore be more important to these offspring. If it is true that each kind of mothering has different advantages, or has advantages at different infant and juvenile ages, then we must also consider that infants of high-ranking females are not as subject to the socially induced dangers of laissez-faire mothering as are infants of low-ranking mothers. Thus by this argument, high-ranking mothers can in fact "afford" to be more laissez-faire; that is, laissez-faire mothering might be good mothering for a high-ranking infant, bad for a low-ranking one, and vice versa for restrictive mothering.

Many of the attempts to evaluate parental care in different Western social classes ignore the fact that not only the options that parents have but also the outcome situations to which infants must adapt may differ among these groups (see review in Kohn 1977). In contrast, anthropological studies of child rearing (e.g., LeVine 1973, Whiting and Whiting 1975, Whiting and Child 1953, and the review of socialization by Draper 1974) often consider outcomes for the particular social organization as a major determinant of child-rearing practices. If infant survival and adult adjustment, survival, or reproduction are reasonable criteria for good parenting, then the particular parental behavior that constitutes good parenting may vary considerably even within a given human or monkey society and must be considered within its normative context.

We have seen the limitations placed on mothers and infants by their social and physical world, but we have also seen hints of a wide range of possible modes of action within those constraints. Why, I wonder, do not low-ranking baboon females make more use of male associates? Why do some individuals seem to be able to stay away from trouble through use of space, while others seem to get caught in the middle of every fight? In particular, it appeared that among low-ranking females, older ones such as Este and Judy, as compared with Brush and Handle, were better at making use of male associates and of space. Why do some mothers such as Preg and Slinky invest for months, to their own peril, in obviously doomed infants while others such as Gin provide so little care and that so reluctantly?

What are the causes and consequences of some adult females, regardless of age, reproductive status, or dominance rank, being much more actively interested in or aggressive in their interest in new infants? And what of juveniles such as Janet, who stayed near and interacted frequently with mothers, while others, such as Dotty, did so appreciably less? We now know that Janet was still three years from menarche, Dotty only two, facts that perhaps make the differences in their behavior even more surprising. Will Janet be more competent with her first infant than Dotty will be with hers? In laboratory studies, only extremes of social deprivation during infancy have been shown to result in poor mothering during adulthood (see review in Kaufman 1974).

Will restrictive mothers continue to be so with successive infants, and will their daughters be restrictive when they become mothers? We would predict considerable consistency from the persistence of dominance rank differences, but these would be mitigated during adulthood as mothers learned to make better use of resources such as male associates and as maternal aging put a premium for the infant on early independence.

The Period of Semi-Independence

When her infant is in its fourth or fifth month, a mother's social life is beginning to return to preparturition levels. Her infant spends much time out of contact, and some of those who wish to interact with it do so during those times. By this age the infant readily recognizes individuals and often avoids some, seeks out others. The mother's social life decreases rapidly just as she needs to feed much more to provide for this growing infant: as one set of pressures, the social ones, decreases, the ecological ones increase. Stress probably peaks for high-ranking mothers when their infants are about five months old. For low-ranking mothers, for whom social stress is great during the early months, pressures probably remain high for the whole first five or six months. Experiences in these first six months probably provide the main mechanism by which offspring assume their mother's dominance rank.

By the time infants are 6 months old, all have experienced some maternal punishment and have displayed their distress with tantrums. But in response they have restructured their time with their mothers. Also, they have play groups to join, they are agile, well coordinated, and fast. If the season is right they can obtain many foods for themselves, a fortunate thing, too, because they now weigh over 15 percent of their mothers' body weight instead of the 6 to 8 percent they did at birth. If their infants are healthy and weaning foods are abundant, mothers' lives can really begin to return to "normal" when their infants are in their second six months of life. Some mothers resume their menstrual cycles during this period and by the time their offspring are 18 months old, almost all mothers are again pregnant.

What of a mother's semi-independent offspring between 6 and 18 months of age? Increasingly, it is involved in play and agonistic interactions. Less often will its mother interfere when these interactions result in distress for the infant. It is at this age that older siblings and adult male associates may become increasingly important. I think that it is in the second year of life, when mortality rates are still high, that parent-offspring conflict of interest is most likely to occur and that other group members potentially provide the greatest contribution to the youngster's survival.

Demographic Influences

Our ignorance of the 6- to 18-month period constitutes a major gap in our understanding of baboon life histories. It should partially be filled by the current work of Stuart Altmann at Amboseli and of Nancy Nicolson at Gilgil. However, at this time it is worth considering which individuals potentially constitute the world of the older infant and young juvenile. Who these individuals are will vary, determined to a considerable extent by the demographic factors operating in a population. Most long-term primate field studies have been conducted with expanding populations, usually ones with low rates of mortality, particularly for infants; and in the case of provisioned groups, there may also be a higher rate of conception. What is of interest here is the group structure that results from various combinations of age-specific birth and death rates and migration patterns (S. Altmann and J. Altmann 1979), because each produces a particular group composition and each might be comparable to particular conditions of various human groups. High birth rates and low rates of infant mortality, for example, result in large cohorts of immatures. In such groups we might expect to find the tendencies for play groups to segregate by age and sex, as has been observed in laboratory groups of other primates, and might be comparable to the choices human youngsters have in large communities. Alternatively, in a nonexpanding population with appreciable infant mortality such as that at Amboseli, infants will not have accessible a sufficient variety of playmates to select among them. For very young baboon infants, proximity of their mothers (due perhaps to subgroup membership or dominance rank, as in this study) and closeness of infants in age may be more important variables than infant gender. Later, gender can be increasingly important as cohorts consist of wider age ranges and as the youngsters' play becomes less dependent on their mothers' location. Both group size and composition will affect an infant's options. For human children, Draper (1976) and Blurton-Jones and Konner (1973) have discussed similar issues as determinants of play group composition, choice of games, and gender differences in play. These authors contrast play among !Kung children who grew up under demographic conditions similar to those at Amboseli with play of children who live in Western cities with large peer populations. Draper also discusses other factors of social organization and mode of production as they interact with demography and affect children's associations (Draper 1976). Likewise, Barker and Gump (1964) make some of the same comparisons for small towns and large urban or suburban environments.

We know from studies of both human infants and caged primates (e.g., Freedman 1974 and references therein, Jensen et al. 1967, Mitchell 1968, Rosenblum 1974; but see also Blurton-Jones and Konner 1973, Young and Bramblett 1977) that some small gender differences can be detected at very early ages. Yet it appears that under some conditions such differences may be minor or undetectable in ordinary mother and infant behavior and under others become quite important and exaggerated. The former seems to have been the case in the present study, in which other factors swamped any gender differences that might have existed. Will these early experiences just delay emergence of some sex differences or will the modified early experiences in turn modify aspects of adult behavior?

Other ways in which demographic parameters will determine the social milieu and the options available to individuals include whether individuals can surround themselves primarily with kin or not. If mortality is low, mothers can often select kin as associates. Berman (1978) has pointed to this factor as potentially explaining some of the differences in maternal restrictiveness observed in various studies of rhesus monkeys. Utilizing Rosenblum's (1974) framework for classifying and comparing effects of rearing conditions in terms of the complexity of those conditions, Berman observes that Rosenblum assumes that situations that provide more complexity also provide more stress. Berman suggests that although "complexity" is hard to define, it does seem that often complexity does not covary with stress. Thus an elaborate kin-organized group may be considered more complex but may provide less stress for a mother. Berman suggests that we may not be dealing with a single dimension. However, I think that the main issue is often one of level or unit of analysis rather than of dimensionality. Mapping from demography to social organization and from either of these to individual experience must be done with care. A problem occurs when we use a phenomenon defined at the level of the group to predict one at the individual level: as Berman indicates, a complex group social structure may in fact provide opportunities for individuals to have quite structured, simplified lives within a subset of the organization. Difficulty also occurs when we relate immediate consequences of a behavior, its lifetime consequences, and its intergenerational consequences due to either cultural or genetic transmission. We need to use the appropriate level of analysis rather than add dimensions.

Future Research

Perhaps well-designed field and laboratory investigations can enable us to delineate better the role of maternal nutrition, infant nutrition, conspecifics, and time budget constraints in affecting mothers and infants and their relationships. Some variables that are particularly difficult to isolate in a field situation might be investigated through use of experimental procedures, for example, ones in which laboratory animals would be required to spend time and energy gaining food, where maternal nutrition levels would be high but no weaning foods would be available for infants, where low-ranking animals could "buy" some time or space away from threatening peers. More information regarding kin relations and past histories of animals will aid in interpreting results of laboratory studies and in making comparisons between studies.

But any effort at isolating variables will always have to return to the questions: In what range of each variable does a human or other free-living animal usually find itself and with what combination of values of different variables is it usually faced? How do these variables interact? In the real world mothers and infants are not faced with one variable at a time. The effect of any variable is dependent on the values of others, often even in nonmonotonic and discontinuous ways. Although the specific answers to these questions will vary with the species, and with the habitat, I believe that for understanding both human and nonhuman mothers many of the important questions, some of the major variables, and the fruitful research strategies are the same. The recent series of studies of the !Kung hunter-gatherers by Richard Lee, Irven DeVore, and their colleagues exemplifies the potential richness of this approach (e.g., Howell 1979, Lee 1978, Lee and DeVore 1976).

Despite the potential pitfalls in a holistic approach, there will always be limits and distortions to a reductionist or a single-dimensional world-view. We share with baboons a complex existence that is more than the sum of its parts. To understand that totality we will need to integrate results both from experimental strategies that focus a magnifying glass on a few individual points along one or two lines and from field explorations of the full complexity of that existence.

Appendix 1
Maternal Genealogies in Alto's Group

THE FEMALE FROM WHOM an infant or yearling was nursing when the studies began in 1971 is considered that infant's mother, because we have had no instances of adoption or of a baboon infant (that is, an offspring less than 12 months old) surviving its mother's death. The one putative mother-juvenile offspring relationship (that between Ring and Fem) from July 1971 is indicated with a dotted line. Relationships are drawn as in anthropological kinship drawings (e.g., as in Fox 1967). Triangles are used for males, circles for females, squares for a few stillbirths for which sex was not determined. The individual's name appears near the symbol. A line through the symbol indicates that the individual has died. Stillborn infants and some who died shortly after birth were not named. Dates of birth (b) and death (d) are indicated. A darkened symbol indicates that the infant was a subject in the present mother-infant study.

Update May 1979: The following individuals included in the genealogies disappeared in early November 1978: Fem with infant Floret, Este with infant Emet and adult son Toto, and Mom with infant Mario. In January 1979 Dogo was killed by hyenas. Handle gave birth to female infant Hulk in February 1979.

195

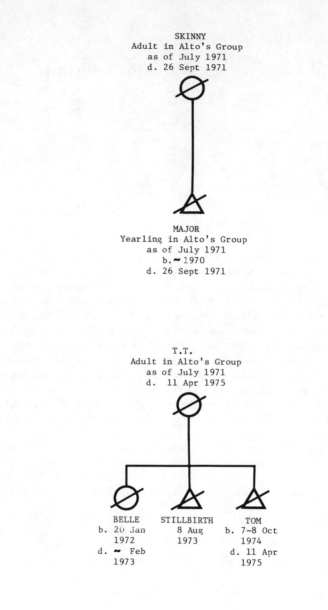

SKINNY
Adult in Alto's Group
as of July 1971
d. 26 Sept 1971

MAJOR
Yearling in Alto's Group
as of July 1971
b. ~ 1970
d. 26 Sept 1971

T.T.
Adult in Alto's Group
as of July 1971
d. 11 Apr 1975

BELLE	STILLBIRTH	TOM
b. 20 Jan	8 Aug	b. 7-8 Oct
1972	1973	1974
d. ~ Feb		d. 11 Apr
1973		1975

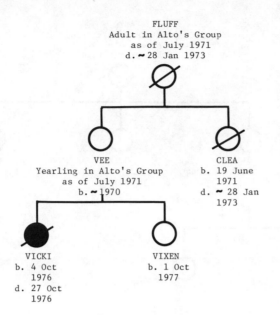

FLUFF
Adult in Alto's Group
as of July 1971
d. ~ 28 Jan 1973

VEE
Yearling in Alto's Group
as of July 1971
b. ~ 1970

CLEA
b. 19 June
1971
d. ~ 28 Jan
1973

VICKI
b. 4 Oct
1976
d. 27 Oct
1976

VIXEN
b. 1 Oct
1977

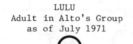

LULU
Adult in Alto's Group
as of July 1971

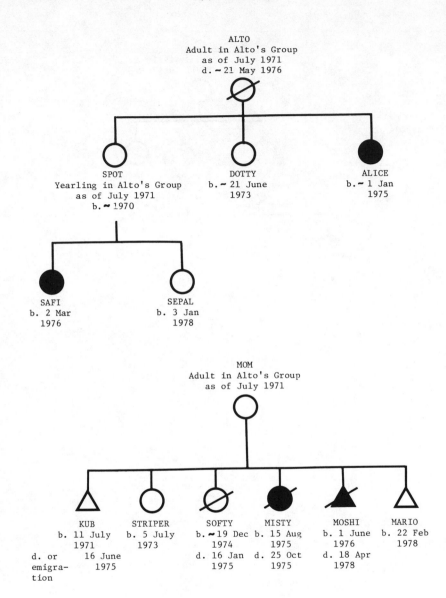

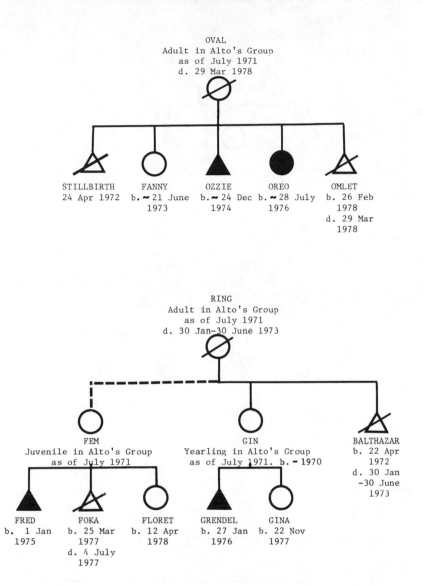

OVAL
Adult in Alto's Group
as of July 1971
d. 29 Mar 1978

STILLBIRTH
24 Apr 1972

FANNY
b. ~ 21 June
1973

OZZIE
b. ~ 24 Dec
1974

OREO
b. ~ 28 July
1976

OMLET
b. 26 Feb
1978
d. 29 Mar
1978

RING
Adult in Alto's Group
as of July 1971
d. 30 Jan–30 June 1973

FEM
Juvenile in Alto's Group
as of July 1971

GIN
Yearling in Alto's Group
as of July 1971. b. ~ 1970

BALTHAZAR
b. 22 Apr
1972
d. 30 Jan
-30 June
1973

FRED
b. 1 Jan
1975

FOKA
b. 25 Mar
1977
d. 4 July
1977

FLORET
b. 12 Apr
1978

GRENDEL
b. 27 Jan
1976

GINA
b. 22 Nov
1977

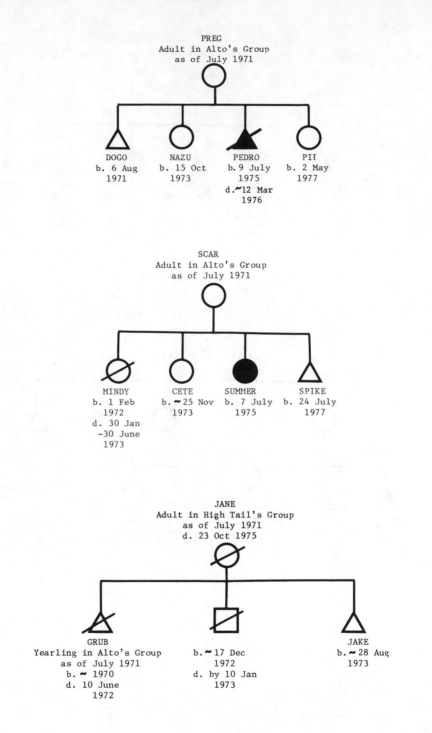

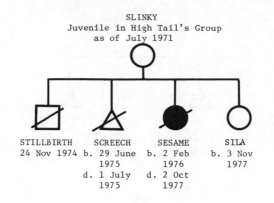

SLINKY
Juvenile in High Tail's Group
as of July 1971

STILLBIRTH	SCREECH	SESAME	SILA
24 Nov 1974	b. 29 June 1975	b. 2 Feb 1976	b. 3 Nov 1977
	d. 1 July 1975	d. 2 Oct 1977	

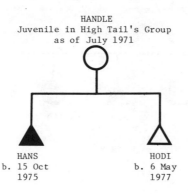

HANDLE
Juvenile in High Tail's Group
as of July 1971

HANS
b. 15 Oct
1975

HODI
b. 6 May
1977

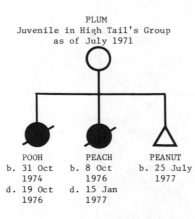

PLUM
Juvenile in High Tail's Group
as of July 1971

POOH	PEACH	PEANUT
b. 31 Oct 1974	b. 8 Oct 1976	b. 25 July 1977
d. 19 Oct 1976	d. 15 Jan 1977	

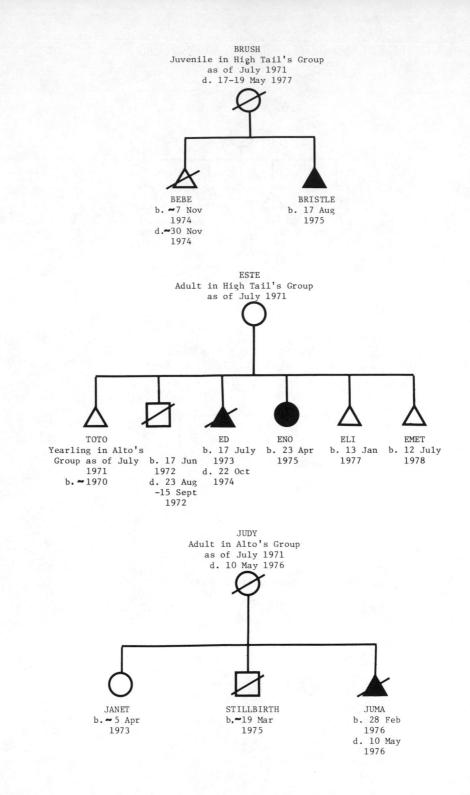

BRUSH
Juvenile in High Tail's Group
as of July 1971
d. 17-19 May 1977

BEBE
b. ~7 Nov
1974
d.~30 Nov
1974

BRISTLE
b. 17 Aug
1975

ESTE
Adult in High Tail's Group
as of July 1971

TOTO
Yearling in Alto's
Group as of July
1971
b. ~1970

b. 17 Jun
1972
d. 23 Aug
-15 Sept
1972

ED
b. 17 July
1973
d. 22 Oct
1974

ENO
b. 23 Apr
1975

ELI
b. 13 Jan
1977

EMET
b. 12 July
1978

JUDY
Adult in Alto's Group
as of July 1971
d. 10 May 1976

JANET
b. ~5 Apr
1973

STILLBIRTH
b.~19 Mar
1975

JUMA
b. 28 Feb
1976
d. 10 May
1976

Appendix 2

Selective Case History Descriptions of All Mother-Infant Dyads with Emphasis on Adult Male and Kin Associations

The descriptions are based partially on ad libitum notes. Infants are ordered by age starting with the oldest. See also Appendix 1.

1. Female Pooh

Plum was very protective of her first infant, Pooh, who was small and exhibited locomotor disability when first observed to leave her mother (J. Scott personal communication). There is no information on any early associations. When I first observed Pooh at the age of eight months, she was thin, smaller than the six-month-olds, retarded in her skin color maturation (J. Altmann et al. 1977), and had a short, scraggy coat and impaired locomotion. She was undergoing severe weaning with frequent distress vocalizations and occasional biting by her mother, who otherwise usually ignored Pooh. At the time Pooh was associated with male Chip, who sometimes carried Pooh, stayed behind the group with her, slept near or with her in the trees and fed near her.

The day after Chip migrated to another group, when Pooh was 15 months old, Pooh appeared with what seemed to be a dislocated hip and could walk only very slowly and clumsily. Male Max immediately assumed Chip's role as "godfather," staying considerable distances behind the group with Pooh, especially during the first few days. Pooh's condition eventually improved to its earlier state. Her relationship with Max continued. One day when Pooh was 24 months old, male Even ran off with Pooh in his mouth, inflicting fatal wounds on her as he did so. Max chased Even until Even dropped Pooh. Max then stayed with Pooh, grooming her and sitting near her. Plum, with her new infant, Peach, ignored these interactions. When the group moved out from the sleeping trees without Pooh the next morning, Max stayed at the rear of the group, repeatedly climbed into trees and faced back toward the sleeping grove while giving barks (at least 26 times between 0815 and 1116) that are characteristic of lost animals (see Stolz and Saayman 1970). The next day Pooh appeared by a shrub near Favorite Grove (J. Walters personal communication). Max approached and sat near her, once embracing her. Max lagged behind as the group moved off (owing to Maasai and their cattle), and as Pooh weakly attempted to follow, moving only a few meters. Cattle separated both Max and Pooh from the rest of the group. Eventually Max rejoined the group after barking and looking back toward where Pooh was last seen. She was never seen again.

2. Male Ozzie

When Oval's healthy male infant, Ozzie, was born, her daughter, Fanny, was one-and-a-half years old. High Tail and Slim were Oval's consorts the week Ozzie was conceived (G. Hausfater personal communication). No information is available on early adult male attachments, but at six months of age Ozzie was often associated with High Tail. Seven months later, High Tail and Slim were Oval's consorts during the period in which she became pregnant with her next infant, Oreo.

3. Female Alice

Alice, a healthy female infant, was born to the oldest and highest-ranking female, Alto. Although B.J., Crest, and Peter each mated with Alto during the week that Alice was conceived, Slim was Alto's consort during the days of most likely conception and was, therefore, probably Alice's father (G. Hausfater personal communication). Alto had two other known living offspring in the group, five-year-old Spot and one-and-a-half-year-old Dotty. At 6 months of age Alice was associated with adult male Peter. An association between Alto and Peter has existed at least since 1971; during the present study an association existed between Dotty and Peter but not between Spot and Peter. In fact, Peter had grooming relations with the other family members, was often near them, even in the sleeping trees, and he supported the younger sisters and Alto in agonistic encounters with Spot. Alto died when Alice was 17 months old, at which age she was still sleeping with and occasionally suckling from her mother. During the next few months, Alice slept in contact with Peter or her sister, Dotty. Alice's association with Peter persisted until his disappearance during 1978 (D. Stein personal communication).

4. Male Fred

An early association was observed between male Max and Fem with her first infant, Fred (D. Post and J. Scott personal communication). This strong, persistent relationship existed well into Fred's second year of life, with Fred often taking advantage of it, feeding immediately next to Max. In this way Fred frequently obtained scraps of mammalian prey when others of his age did not. Max was most probably Fred's father, but B.J. was also a consort during the last week of estrus (G. Hausfater personal communication).

5. Female Eno

Eno's next-surviving sibling was five-year-old male Toto, who was not often associated with Eno or with his mother, Este, during this year. However, Eno and Toto were associated during 1976-77 (J. Walters personal communication and in preparation). When Eno was first observed by me when she was two-and-a-half months old, only a slight association could be detected between her and male B.J., on whom she occasionally rode. No stronger relationship developed before B.J. disappeared from the group nine months later. Before B.J.'s disappearance he and High Tail were Este's most common consorts when she resumed cycling.

6. Female Summer

Both Stubby and Slim stayed near mid-ranking female Scar when her female infant Summer was born, 17 months after Scar's previous infant, Cete. However, Scar often avoided both males and neither she nor Summer developed an association with any other male. Perhaps because Summer and her mother remained in general rather isolated from other adults, Summer stayed in unusually close proximity to her mother as well as spending much time in actual contact. Cete associated with her younger sister only after Summer became independent 6 to 8 months later.

7. Male Pedro

Peter, Slim, and briefly Stubby were associated with mid-ranking older female Preg when her male infant Pedro was born. When Pedro was four months old, Stubby died after suffering a severe leg wound, probably inflicted by Ben. As with Slim's other such relationships that year, he did not carry the infant and soon lost interest in the pair. Uncharacteristically, Peter did so as well. Preg's two-year-old daughter, Nazu, was not associated with the pair; her four-year-old son, Dogo, was with them only occasionally. By three months of age Pedro's locomotion was decidedly abnormal and his coat and skin color development were retarded (Fig. 42). He deteriorated rapidly in the sixth month. At that time, Nazu began to associate with him and groom him. She continued to do so until his death at eight months of age. Slim was Preg's sole consort when she resumed cycling.

8. Female Misty

Slim and Peter were frequent neighbors in the period immediately after female infant Misty was born to Mom, an elderly, high-ranking female. B.J., Chip, and Slim had been Mom's consorts during the week Misty was conceived (G. Hausfater personal communication). No relationship developed between Misty and either male before Misty's death at two-and-a-half months of age. Partly because Slim threatened Striper away, Misty's two-year-old sister, Striper, was only slightly associated with her and her mother during Misty's brief life. Striper was more closely associated with Mom when Mom's next infant, Moshi, was born. Slim was Mom's sole consort when she resumed cycling after Misty's death.

9. Male Bristle

Brush's first infant was born with webbed fingers and died in the first two weeks of life (D. Post and J. Scott personal communication). Thereafter, Brush became pregnant in one cycle. Ben and Max both attempted to intensify their existing relationships with Brush when male infant Bristle was born six months later. Ben displaced Max and thereafter established one of the strongest and most persistent male-infant relationships observed, interrupted only for several months during which Ben suffered a severe shoulder wound, probably inflicted by male Red.

10. Male Hans

Handle was, along with Slinky, one of the two most protective and restrictive mothers when her first infant, Hans, was born. When Hans was two days old he was kidnapped by adult female Gin and kept until he was dehydrated and unable to cling; finally, he was successfully retrieved by his mother the next morning. Like other primiparous females, Handle took more than a month to accept and reciprocate or make use of a male's attempts to remain nearby, the male in her case being Even. Once established, however, the relationship between Handle and Even was one of the most reciprocal, and often Handle, Hans, Even, and Even's consort female, Lulu, constituted a close subgroup unit. Red, who also attempted association with Hans and Handle in the first month, was displaced by Even and was repeatedly avoided by Handle, perhaps because Red had supported Gin when she avoided returning Hans after the kidnapping.

11. Male Grendel

Gin's association with young male Red persisted into male infant Grendel's first month. Gin was extremely rejecting and punitive with this first infant, and a relationship between Grendel and Red began to develop. At this time, Red was undergoing a period of rapid rise in rank and was frequently involved in agonistic interactions with other males, during which he probably inflicted Ben's serious wound (section 9, above). These interactions culminated in Red's leaving the group, at first for two months, then sporadically thereafter. No relationship with Grendel was reestablished. Rather, during month two Grendel established a relationship with male High Tail, who had no obvious association with Gin at the time. High Tail was the only potential father recorded for Grendel but we have incomplete consort data for the cycle in question. Grendel rode on High Tail often and for long periods (Fig. 36), fed next to High Tail and sat in his shade on hot afternoons, ran to him in times of alarm or other distress, until High Tail was killed by a leopard when Grendel was four months old. Thereafter, Grendel spent several weeks attempting to ride on many other individuals, and also tried again to spend more time in contact with his mother, but his mother remained rejecting and punitive. Most group members will readily carry only black infants; they only allowed Grendel to ride on them briefly. Grendel survived this difficult period, soon adjusting to his greater independence.

12. Female Sesame

Slinky's first pregnancy resulted in stillbirth, her second in a severely defective infant, who neither clung nor suckled, and who died two days after birth. Her third infant, Sesame, seemed weak from birth, but survived. Sesame's locomotion was poor and her maturation retarded. When she was four months old she suffered a period of illness characterized by severe locomotor problems and perhaps partial paralysis. After Sesame's birth, Max and Ben both sought association with highly protective Slinky, Max rapidly displacing the recently wounded Ben, whom Slinky often avoided. In the first month, Even stayed nearby but was noninteractive. Slinky neither approached nor avoided him and he lost interest in the pair thereafter. Max remained moderately associated with the pair, but shifted much of his attention to Pooh when Chip emigrated. Either Max or Even was probably Sesame's father.

Sesame disappeared and was presumed dead during the second year of life.

13. Male Juma

A good example of the dynamics of the male associations is the case of Judy and her infant, Juma. We have only partial consort data for the cycle in which Judy became pregnant, but her frequent associate in general was male Stubby, who died early in the present study, shortly after Judy became pregnant. During late pregnancy Judy was mildly associated with male Peter. Male Slim persistently followed Judy when Juma was born and even more persistently followed infant-snatching Spot, who, with her own new infant, followed Judy. Spot sometimes groomed and stayed near Slim. Judy repeatedly moved away from both of them, but was often in their presence, primarily because they persistently followed her. Peter stayed somewhat farther away at these times. Peter and Judy did approach each other. In contrast to the way she reacted to Slim, Judy appeared calm near Peter, did not restrain her infant or avoid Peter, and did groom him. Three-year-old Janet was often nearby and engaged in grooming relations with her mother and younger brother.

When Juma was two-and-a-half months old, Judy and Juma disappeared during the night following the first day of what appeared to be a virus epidemic in the group. Peter was sick that day. He was one of the last two baboons to leave the sleeping grove, and as he slowly joined the group he repeatedly stopped and looked back at the grove. He was the only one to do so. We could detect no reaction to Judy and Juma's disappearance either by Slim or by Janet.

14. Female Safi

During early 1974–75, Spot had been associated with male Stubby (D. Post personal communication), who died soon after Spot became pregnant. On the cycle in which she became pregnant her only consort recorded on days of likely conception was Stubby, but data are not available for all such days.

When Spot gave birth to her first infant, female Safi, Spot ranked second only to her mother (Alto) in the female dominance hierarchy. Only her yearling sister, Alice, and male Slim intruded upon Spot and her infant. Slim was virtually the only animal who pulled Safi. Alice frequently pushed into Spot's ventrum, as she sometimes did with the other mothers, sitting on or pushing the infant Safi, actions that Spot tolerated until Alice became too persistent and was pushed away. Alice sometimes then enlisted support from her mother, Alto, from adult male Peter, or from her three-year-old sister, Dotty. At an early age Safi developed considerable independence.

During the apparently viral epidemic in May 1976, Spot and her infant, Safi, then seven weeks old, were two of the sickest baboons in the group. During the first few days of their illness, characterized by minimal, swaying movement, long rests, and no feeding, juveniles and low-ranking females who had not ventured to do so before approached and handled the infant. There seemed to be a clear recognition of Spot's incapacity, yet Spot did not drop in dominance rank during her illness, that is, there were no reversals in agonistic outcomes. Spot recovered more quickly than Safi, who retained an appreciable limp for a month and never resumed dorsal riding.

15. Male Moshi

Moshi, a healthy male infant, took full advantage of Mom's nonrestrictive care and the relative lack of interference from others. He rapidly became one of the most independent infants. Moshi's sister, three-year-old Striper, spent considerable time with her mother and new brother. Slim, Mom's only consort during the cycle of conception, was almost surely Moshi's father. Through repeated threats Slim displaced Peter as an associate of the mother and infant. The subsequent association was mild, did not last long, and included Slim's usual infant pulling. When he was about eight to ten months old, Moshi's coat turned white (J. Walters personal communication), a sign of poor health that usually appears about the fourth month (see Chapter 8; also Pedro and Sesame in this appendix). Moshi fell from a tree and died during his second year of life (D. Stein personal communication).

16. Female Oreo

When Oreo was born, one of the two males likely to have been her father, High Tail, was dead. The other, Slim, was early associated with the mother-infant pair (G. Hausfater, personal communication). Little evidence of such an association remained when I first observed Oreo, during her third month. During the last trimester of Oval's pregnancy, all ties seemed to be at least temporarily severed between her and her previous offspring, Ozzie, a state which had not changed four months later. A close association did exist between Oval and her three-year-old daughter, Fanny, as did one between Fanny and Ozzie.

17. Female Vicki

Vee's first infant, Vicki, was not able to get on the nipple during her first day of life; her mother carried her upside down (ventrum down) and backward, even dragging her and bumping her on the ground much of the first day. The infant seemed normal at first sight in the morning, but the extreme mishandling she received soon made it hard to tell the source of poor coloration and signs of weakness and dehydration observed later that first day. In the sleeping trees the next morning, the infant was on the nipple when first seen. Vee behaved much more competently but remained fairly unresponsive and quite unprotective of the infant. She was still incompetent compared even with the other primiparous females. Vicki's early deprivation is the most likely cause of her death at three weeks. She never regained adequate skin color or clinging ability and often looked somewhat emaciated. During these few weeks, Vee was associated with male Slim, most likely Vicki's father.

18. Female Peach

As the group foraged, Plum remained behind to give birth to her second infant, daughter Peach, after about one to one-and-a-half hours of labor. At the onset, she handled and then licked the vaginal fluids. Increasingly, she thrust her arm in a sudden jerky motion or grasped and strained against a large log, grimacing. She was restless and attentive throughout. Just as the infant's head began to appear, a Maasai tribesman walked by in the distance and Plum moved under a shrub. She reappeared with infant Peach after 20 minutes and very rapidly ate the afterbirth, making several gagging motions. She finished the placenta and bit the end of the umbilicus.

For almost two hours, Max remained about 100 yards from Plum, high on a fallen, dead tree, watching her and occasionally giving single, two-phased alarm barks. Peach's sister, Pooh, remained near Max. Max, Pooh, Plum, and Peach rejoined the group, staying near each other the rest of the afternoon. The next morning, Red kept Max away from Plum. Male Even then displaced and chased Red away, remaining in a close reciprocal relationship with Plum and Peach throughout the remainder of the study (three weeks). Plum, wary of others and quite protective of her infant, followed Even closely. Even was most likely to have been Peach's father. Red, Max, and High Tail (dead by Peach's birth) were the other consorts during Plum's last estrous period. At three months of age, Peach became sick and died of unknown causes.

Appendix 3
Selective Case History Descriptions
of All Adult Males

FOURTEEN MALES WERE ADULTS in Alto's Group during some part of the 1975–76 study year, including Stiff, Red, Russ, and Stu, who matured to adulthood in the group during that year. Their relative dominance rank during June and July of 1975 and changes in relative dominance relationships during the subsequent year are indicated in Table 19. Other changes in the overall hierarchy occurred because of deaths and migrations, as indicated previously in Table 1. Below is an alphabetical listing of the males and a brief summary of each male's history (see also Appendix 2). G. Hausfater, F. Saigilo, D. Stein, and J. Walters have contributed demographic data from mid-1976 through 1978.

1. Ben

Ben was a juvenile male when studies began in July 1971. During the summer of 1973 he was ninth- or tenth-ranking in the adult and subadult male hierarchy, twelfth-ranking in 1974, and during 1974–75 he rose to second rank. We suspect that he inflicted male Stubby's fatal wound in October of 1975. In February of 1976, he received a similar canine puncture wound, probably inflicted by Red. He dropped in rank (see Table 19). His recovery was extremely slow, impeded by infection, but by fall of 1976 he was beginning to take part in social interactions again, he stopped limping in early 1977, and in December 1978 I could detect no limp or other residual signs of the wound. During 1975–76 he was closely associated with Brush and Bristle (see Appendix 2), a relationship still detectable in 1976–77 (J. Walters, personal communication) and in 1978 (D. Stein, personal communication).

2. B.J.

B.J. was an adult male in Alto's Group in July 1971. In August he was seriously wounded, probably in a fight with the dominant male, Ivan, who immediately took over B.J.'s consort female. B.J. left the group, lived alone for a period, then joined another group temporarily (Hausfater 1975a). He eventually rejoined Alto's group and became the dominant male of the group for

Table 19. Adult male dominance ranks.

Dominance hierarchy for males 1 July 1975		Changes in dominance hierarchy from 1 July 1975–31 Oct. 1976	
Male	Rank	Date	Change
Slim	1		
Even	2		
Stubby	3		
Ben	4		
Max	5		
Peter	6		
High Tail	7		
B.J.	8		
Chip	9		
Dutch	10		
Stiff	11		
Red	12		
Russ	13		
Stu	14		
		11 Aug. 1975	Red over Dutch
		29 Aug. 1975	Ben over Stubby
		1 Oct. 1975	Red over B.J.
		1 Oct. 1975	Red over Max
		9 Dec. 1975	B.J. over Max
		11 Dec. 1975	Russ over Max
		25 Dec. 1975	B.J. over High Tail
		18 Feb. 1976	Red over High Tail
		27 Feb. 1976	Red over Ben
		8 Apr. 1976	Russ over Peter
		3 May 1976	High Tail over Max
		22 May 1976	Peter over Ben
		24 May 1976	Max over Ben
		1 June 1976	Stu over Peter
		1 June 1976	Stu over Max
		12 June 1976	Stu over Ben

one to two years (Slatkin and Hausfater 1976). He was mid- to low-ranking thereafter. He disappeared from Alto's group in March 1976, and we have not been able to locate him since then.

3. Chip

Chip joined Alto's Group as a fully adult male in August of 1973. He was never very high-ranking, but remained in Alto's Group and during 1975 was associated with Plum and her offspring Pooh. He disappeared from the group in March 1976, was tentatively identified in Kit South Group soon thereafter, but was not seen there during 1976–1978.

4. Dutch

Dutch was a fully adult male of Alto's Group in July 1971. Although low-ranking, he was frequently in consort with adult females (Hausfater 1975a). During 1974 he developed a skin ailment on his hands that greatly impaired his locomotion and resulted in his walking on his knuckles (Hausfater 1975b). During early September 1975 he developed a severe respiratory disease, appeared to be feverish, and was unable to follow the group. The group encountered him several times in their travels over the next week. He looked thinner and more ill each time. He was last seen by us on 6 September 1975. We assume he died shortly thereafter.

5. Even

Even was a juvenile in Alto's Group in July 1971. During 1974–75 he rose rapidly in rank and he frequently migrated in and out of Alto's group (D. Post, personal communication), often chased for long distances from the group by Slim. He remained in Alto's Group as second-ranking male for most of the 1975–76 year, during which time he was closely associated with Handle and Hans and then with Plum and Peach. During the latter part of this year he frequently engaged in long chases of the third-ranking male, Red, as Slim had done with him. Red finally became dominant to Even during April 1977 (G. Hausfater personal communication). Even was one of 11 group members (5 adult males, 3 females with infants) who disappeared early in November 1978. Despite extensive searching and censusing of all other groups, they could not be located anywhere in the Amboseli area (D. Stein and J. Walters personal communication) and were presumed dead.

6. High Tail

High Tail was the only adult male of the small group that fused with Alto's Group in fall 1972 (J. Altmann et al. 1977, McCuskey 1975). At that time he became the dominant male of Alto's Group. He was mid-ranking during 1975–76 and was closely associated with Gin's son, Grendel, in that year (see Appendix 2). He became quite ill during the presumed viral epidemic of May 1976. Just as he was recovering, he was apparently attacked by a leopard during the early morning hours of 16 May: we found him, partially eaten, in a tree that morning with the leopard nearby and the whole baboon group in an agitated state. When they made a late descent and moved compactly and rapidly away from the sleeping grove, several juveniles looked up into the nearby tree where High Tail's body hung, behavior that led us to the corpse.

7. Max

Max was a young adult male in Alto's Group in July 1971. He was fifth-ranking in July 1975 and below the newly maturing males in rank during 1975

−76. He protected Pooh after Chip disappeared and rescued her from Even when the latter inflicted Pooh's fatal wounds in October 1976. Max stayed behind the group while Plum was in labor and while she gave birth to Peach (see Appendix 2). In February 1977 he left Alto's Group and joined Hook's Group. He was gone from Hook's Group by July of that year and has not been seen since.

8. Peter

Peter was a fully adult male in Alto's Group in July 1971. He was often associated with Alto at that time and with her and her offspring afterward. During 1976 he was closely associated with Judy when Juma was born (see Appendix 2). When higher-ranking males were in consort with certain females, Peter often counterchased (Hausfater 1975a) the males, screeching, grimacing, and cackling, the outcome of which was that Peter was in consort with the female. Peter was often associated with mothers and infants, including during 1975– 76, was particularly gentle with infants, and often tolerated them very close while he fed on vertebrate prey. Although quite sick during the viral attack of May 1976, he recovered and remained in the group until his disappearance in October 1978.

9. Red

Red was a juvenile in Alto's Group in July 1971. During 1975–76 he was closely associated with female Gin, supporting her when she kidnapped infant Hans, and preventing Handle from retrieving Hans. Perhaps this is why Handle resisted his attempts to associate with her while Hans was small. Red rose rapidly to third rank in the dominance hierarchy during that year (Table 19) and then migrated in and out of the group during 1976 and 1977 before becoming dominant to male Even in April 1977. Red disappeared in November 1978 and was presumed dead (see section 5 above).

10. Russ

Russ was a juvenile in Alto's Group in July 1971. From physical appearance, and less so from behavior, we strongly suspected that he was Alto's offspring. Russ suddenly and rapidly rose in rank in 1976 (Table 19) and migrated from the group that April. When located several times shortly thereafter he was an isolate rather than being in an association with any other group. He was last seen in 1977.

11. Slim

Slim was first identified as an adult male in Stud's Group. He joined Alto's Group sometime between September 1973 and May 1974, at which time he was the dominant male of Alto's Group, and he remained so until mid-1978. During 1975–76 he was the only adult male who frequently pulled infants and in general was quite rough with them. During that year he seemed to lose his interest in each infant after it was a couple of months old, and older infants did not seem to seek his protection or feed near him. Slim disappeared and was presumed dead in November 1978 (see section 5 above).

12. Stiff

Stiff was a juvenile in July 1971. At least since that time he had a limping walk. He was relatively noninteractive, and he had not risen in dominance rank before he left the group in February 1976. For about a month we saw him occasionally, always as an isolate and in the same part of the home range each time.

13. Stu

Stu was a juvenile in Alto's Group in July 1971. In December 1975 and January 1976 he began a rapid rise in dominance (Table 19). He migrated to Stud's Group in 1976 (Table 19), moved to Hook's Group for several months during 1977, and then returned to Stud's Group, where he remained as first- or second-ranking male in December 1978.

14. Stubby

Stubby was a high-ranking male in Alto's Group in July 1971 and remained so until he was fatally wounded, probably by Ben, in October 1975. The puncture wound became badly infected and by mid-November he hardly moved. He disappeared the night of 12 November and was presumed dead.

Appendix 4

Behaviors Recorded in This Study and Analyzed in the Text

AN (s) PRECEDING A BEHAVIOR indicates that it was included in the set of submissive or distress behaviors analyzed in this study: a preceding (a) is used for behaviors of aggression. "A-"followed by a number indicates the comparable behavior in S. Altmann's catalogue for rhesus monkeys, *Macaca mulatta* (S. Altmann 1962); "H-S" and "H-A" followed by a number indicate, respectively, a submissive behavior and an aggressive behavior in Hausfater's study of yellow baboons (Hausfater 1975a). See also Kaufman and Rosenblum (1966), Hinde and Rowell (1962), and Rowell and Hinde (1962) for detailed macaque ethograms including descriptions and some plates.

Approach (A-4): directed walk toward, with orientation to individual rather than to spot; distinguishable by direction of looking and by end-point of approach and by any reorientation if object moves.

Approach-avoid: limbs (especially forelimbs) of animal are flexed, shoulders lowered, chin thrust out, animal looks at object (often infant in mother's ventrum) and, rocks forward and back—forward often while looking at infant, back while tensing and looking (or giving brief repeated "anxious" glances) at mother.

(s)Avert body (included in A-52): directed turn of body away from another, but not with the tensing and twisting of a "cower" or the rigidity of a "cower" or of "fear paralysis."

(s)Avert head (H-S2): turn of head away from; often also with "avert stare"; see also "avert body."

(s)Avert stare, avoid looking at (A-34, H-S2): usually a fixed looking elsewhere than at the individual who is directly in front of the actor's eyes.

(s)Avoid (A-43, H-S4): directed walk away from; identified by its being preceded or accompanied by brief rapid glancing or a very mild cower.

(a)Bite, nip (A-26, H-A13): obvious.

Brush past (probably included in H-A12): walk past, touching, and usually with body swept against, the object.

(a)Chase (A-42, H-A11): rapid run toward with repeated change of direction, tracking the object animal, and not accompanied by gestures of fear or submission.

Clamber, climb on: obvious; usually done by infants.

(s)Coo, cry (A-56): pursed-lip vocalization usually done by infants; hard to locate; usually done in a distress (e.g., weaning) rather than a fright context (I have heard an adult give this vocalization only once—female Handle, when her two-day-old infant was kidnapped and she was unsuccessful in recovering it).

(s)Cower (H-S3; see Hausfater 1975a): "lateral flexion of the spine, often from a seated position"; head often lowered into shoulders; incipient "avoid" —often followed by "avoid'; or incipient "fear paralysis," often followed by that behavior.

Eyelid displays: variety of eyelid and brow movements among which I usually did not distinguish.

(s)Eee (A-36): screech vocalization; occurs in a fear context.

Embrace (infant) (A-12): arm about other animal.

(s)Fear paralysis (H-S5; see Hausfater 1975a): an extreme behavior in which the animal remains rigid, close to the ground; it often follows and grades from "avert stare," "avert head," "avert body," "cower."

Flex arms: arms of animal flexed lowering front half of body and usually with rear in air while looking at another, usually smaller individual (Fig. 43); often accompanied by lipsmacking and cocking of the head; seemed to be a pacificatory and inviting gesture to infants; approach-avoid is this behavior alternated with a tense drawing away and anxious glancing.

Follow (A-45): repeated change of tempo and/or direction of walk with respect to and temporally "in response to" one animal's movements more than to the group or subgroup as a whole.

(a)Grab, hold down (H-A12): obvious.

Groom (A-51): picking through the fur, usually of another animal; one of the most visible, durative, and common primate social behaviors (see Sade 1965 for a detailed description of macaque grooming).

Handle infant: moving or manipulating some part of infant rather than just touching it; but see also "pull (infant)" for that distinction.

(a)Hit, Swat (A-27, H-A12): obvious.

Hold nipple, attempt suck (A-21, A-23): reaching toward the nipple with hand or with mouth but without contact or barely touching; distinguished from "rooting" by hesitancy of movement and by the ability to locate the nipple.

Ignore (A-54): obvious; "it was sometimes obvious that a monkey saw, heard, or felt a behavior pattern that was directed conspicuously toward it, yet 'ignored' the other monkey or monkeys involved" (Altmann 1962).

(s)Ikk (A-24, H-S6): also called gecker or cackle; a vocalization often alternating with "coo"; given in immediate response to, say, a maternal rebuff, then followed by a coo; emitted in a context of low-level fear as well as moderate distress.

Lipsmacking (A-39): rhythmic lip and tongue movements, repeated rapidly.

(s)"Look anxious" (A-35, HS-1): repeated rapid glancing without eye-to-eye contact, usually with some head jerking, tensing.

Look at (A-33): obvious; but not "stare at" or "look anxious."

Lose grasp, lose hold, fall: obvious.

(a)Lunge at (A-28, H-A7): incipient chase.

Mounting (sexual) (A-3, 4, 5, 8): see Altmann 1962 and Hausfater 1975a; non-cycling females do not receive mounts with intromission; only brief pelvis-grabbing, usually in response to a sexual presentation, occurred more than rarely in this study.

Muzzle-muzzle: placement of one animal's muzzle in contact with another's.

Muzzle anogenital area: obvious.

Muzzle other body part (A-53): obvious.

Pause: break in motion, a turn, a look back at another animal; not reliably distinguished from "wait" except a pause is shorter, not held until another joins the actor.

Play (A-25): a gross category; I did not usually distinguish components.

Present for grooming (A-50): animal presents its trunk, broadside; if seated, trunk presented often with arm up and head lifted and tilted.

Present (sexually) (tail up) (A-2): animal presents its anogenital area to muzzle of another animal, usually with one leg back, tail often raised.

Pull (infant) (A-52): when the infant is in contact, and is clinging or is being embraced, this pulling is distinguished from "handling" by the fact that the infant's body is moved toward the actor and away from its present location with sufficient force that present contact would be broken if infant and/or "surface" (e.g., mother) did not exert counteracting force.

(a)Push, shove, lean on: usually a gross body movement or trunk movement rather than a grab, hold down, or hit; but occasionally done with a hand and then distinguishable from a hit by being slower with more contact time and without a grasp.

Restrain infant: actively (hold, embrace, etc.) prevent infant from leaving contact while infant is making movements that would take it out of contact if not restrained; thus this is a dyadically defined behavior.

Ride ventrally (A-13): cling ventrally to the trunk of a standing or walking individual.

Rooting: mouth and head movement, primarily lateral but with some lifting, searching "for" the nipple; only done by very young infants.

(s)Run away from (A-44, H-S4): rapid avoid; see "avoid."

Run toward (A-42): rapid approach; see "approach."

Sit in (or climb in) ventrum of: obvious; usually done by infants.

Soft grunts (A-48): low, rhythmic grunts usually given while animal is approaching or seated looking at an infant.

(a)Stare at (A-32, H-A1): prolonged, intense, fixed look at.

Suckle (A-19): obvious; in practice I could usually determine only if an infant had the nipple in its mouth and not whether it was actually sucking.

(s)Tail up (A-37, H-S7): tail is held high, often arched, but not in a posterior presentation.

(a)Threaten (includes H-A2-6): gross category lumped in this study; these threats occurred so rarely in my samples that I lumped them; H-A8-10 did not occur at all.

Touch: obvious; but see "handle infant" and "pull (infant)" for distinction.

Wait: pause in walk until the object individual reached the actor, at which time the actor immediately resumed its walk; see also "pause"—not reliably distinguished from pause except by duration and joining by the actor.

Appendix 5

Residuals from Linear Regression of Daily Time Infants Spent in Contact with Their Mothers at Each Age

Residuals are first plotted for all points (each infant on each sample day), then separately for each infant for whom data are available for at least two months and for whom we know the style of mothering (Chapter 7). After two weeks of age some infants tend to have positive residuals, that is, are in contact more than expected, others negative. See Chapter 8 for further discussion. See text and Appendix 2 for discussion of illness for Misty, Pedro, Safi, and Sesame.

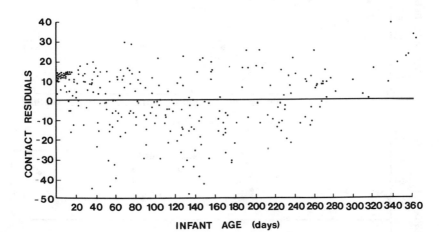

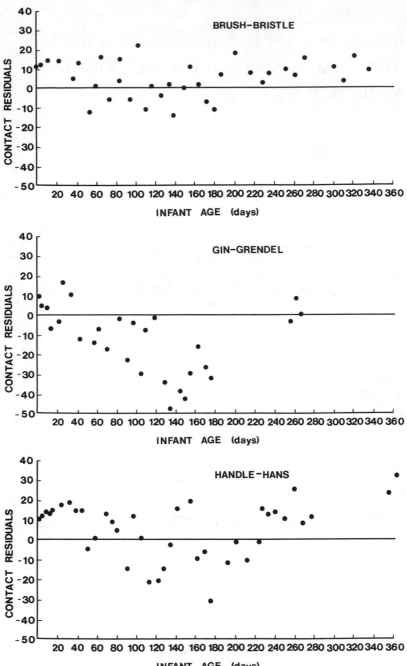

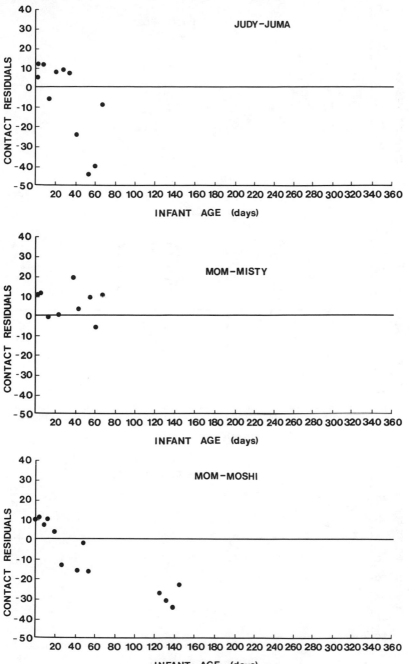

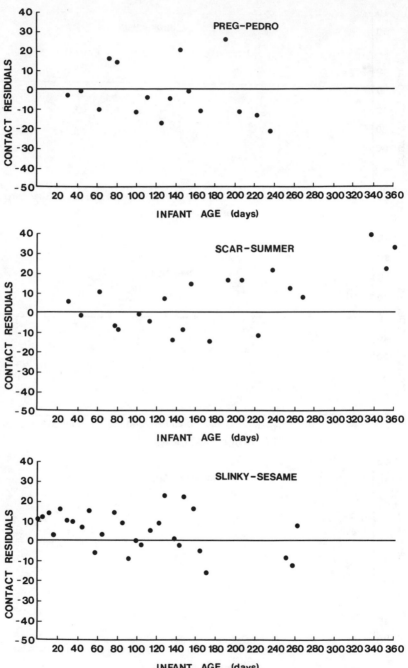

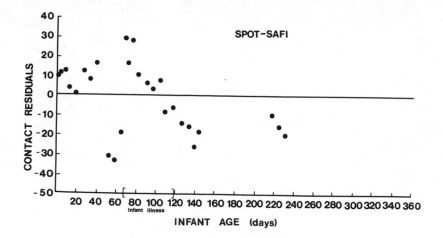

References

ABEGGLEN, H., AND J.-J. ABEGGLEN. 1976. Field observation of a birth in hamadryas baboons. *Folia Primatol.* 26:54–56.

ABEGGLEN, J.-J. 1976. On socialization in hamadryas baboons. Ph.D. thesis, University of Zurich.

ALEXANDER, R. D. 1974. The evolution of social behavior. *Ann. Rev. Ecol. Syst.* 5:325–383.

ALLEE, W. C. 1931. *Animal aggregations: a study in general sociology.* Chicago: University of Chicago Press.

ALTMANN, J. 1974. Observational study of behavior: sampling methods. *Behaviour* 49:227–267.

———— 1978. Infant independence in yellow baboons. In *The development of behavior: comparative and evolutionary aspects,* ed. G. M. Burghardt and M. Bekoff. New York: Garland STPM Press.

———— 1979. Age cohorts as paternal sibships. *Behav. Ecol. Sociobiol.* 6:161–164.

ALTMANN, J., S. A. ALTMANN AND G. HAUSFATER. 1978. Primate infant's effects on mother's future reproduction. *Science* 201:1028–1030.

ALTMANN, J., S. A. ALTMANN, G. HAUSFATER, AND S. A. McCUSKEY. 1977. Life history of yellow baboons: physical development, reproductive parameters, and infant mortality. *Primates* 18:315–330.

ALTMANN, S. A. 1962. A field study of the sociobiology of rhesus monkeys, *Macaca mulatta. Ann. N.Y. Acad. Sci.* 102:338–435.

———— 1970. The pregnancy sign in savannah baboons. *Lab. Anim. Dig.* 6:7–10.

———— 1974. Baboons, space, time, and energy. *Am. Zool.* 14:221–248.

ALTMANN, S. A., AND J. ALTMANN. 1970. *Baboon ecology: African field research.* Chicago: University of Chicago Press.

———— 1979. Demographic constraints on behavior and social organization. In *Ecological influences on social organization,* ed. E. O. Smith and I. S. Bernstein. New York: Garland STPM Press.

ALTMANN, S. A., AND S. S. WAGNER. 1978. A general model of optimal diet. *Rec. Adv. Primatol.* 4:407–414.

ANGST, W., AND D. THOMMEN. 1977. New data and a discussion of infant killing in old world monkeys and apes. *Folia Primatol.* 27:198–227.

ANTHONEY, T. R. 1968. The ontogeny of greeting, grooming, and sexual motor patterns in captive baboons (superspecies *Papio cynocephalus*). *Behaviour* 31:358–372.

BALDWIN, L. A., AND G. TELEKI. 1972. Field research on baboons, drills, and geladas: an historical, geographical, and bibliographical listing. *Primates* 13:427–432.

BARKER, R. G., AND P. V. GUMP. 1964. *Big school, small school.* Stanford: Stanford University Press.

BATESON, G. 1963. The role of somatic change in evolution. In Bateson, *Steps to an ecology of mind.* New York: Ballantine Books.

BECKER, G. S. 1965. A theory of the allocation of time. *Econ. J.* 75:493–517.

BERMAN, C. M. 1978. Social relationships among free-ranging infant rhesus monkeys. Ph.D. thesis, Cambridge University.

BERNSTEIN, I. S. 1976. Dominance, aggression, and reproduction in primate societies. *J. Theor. Biol.* 60:459–472.

BLACKBURN, M. W., AND D. H. CALLOWAY. 1976a. Basal metabolic rate and work energy expenditure of mature pregnant women. *J. Am. Diet. Assoc.* 67:24–28.

———— 1976b. Energy expenditure and consumption of mature, pregnant, and lactating women. *J. Am. Diet. Assoc.* 69:29–37.

BLICK, J. 1977. Selection for traits which lower individual reproduction. *J. Theor. Biol.* 67:597–601.

BLURTON-JONES, N. G., AND M. J. KONNER. 1973. Sex differences in the behavior of Bushman and London two- to five-year-olds. In *Comparative ecology and behavior of primates,* ed. J. Crook and R. Michael. New York: Academic Press.

BOESE, G. K. 1975. Social behavior and ecological considerations of West African baboons (*Papio papio*). In *Socioecology and psychology of primates,* ed. R. H. Tuttle. The Hague: Mouton Publishers.

BOWDEN, D., P. WINTER, AND D. PLOOG. 1967. Pregnancy and delivery behavior in the squirrel monkey, *Saimiri sciureus,* and other primates. *Folia Primatol.* 26:54–56.

BRAMBLETT, C. A. 1969. Non-metric skeletal age changes in the Darajani baboon. *Am. J. Phys. Anthrop.* 30:161–172.

BREUGGEMAN, J. A. 1973. Parental care in a group of free-ranging rhesus monkeys (*Macaca mulatta*). *Folia Primatol.* 20:178–210.

BRODY, S. 1945. *Bioenergetics and growth.* New York: Reinhold.

BUSS, D. H., AND O. M. REED. 1970. Lactation of baboons fed a low protein maintenance diet. *Lab. Anim. Care* 26:709–712.

CAUGHLEY, G. 1966. Mortality patterns in animals. *Ecology* 47:906–918.

CHANCE, M. R. A. 1967. Attention structure as the basis of primate rank orders. *Man* 2:503–518.

CHANCE, M. R. A., AND C. J. JOLLY. 1970. *Social groups of monkeys, apes, and men.* New York: E. P. Dutton.

CHARLESWORTH, B. 1978. Some models of the evolution of altruistic behaviour between siblings. *J. Theor. Biol.* 72:297–319.

CHARNOV, E. L. 1977. An elementary treatment of the genetical theory of kin-selection. *J. Theor. Biol.* 66:541–554.

CHASE, I. D. 1980. Cooperative and non-cooperative behavior in animals. *Am. Nat.* 115

CHENEY, D. L. 1977. The acquisition of rank and the development of reciprocal alliances among free-ranging immature baboons. *Behav. Ecol. Sociobiol.* 2:303–318.

————— 1978. Interactions of immature male and female baboons with adult females. *Anim. Behav.* 26:389–408.

CHISM, J., T. E. ROWELL, AND S. M. RICHARDS. 1978. Daytime births in captive patas monkeys. *Primates* 19:765–768.

CLARK, A. G. 1978. Sex ratio and local resource competition in a prosimian primate. *Science* 201:163–164.

COHEN, J. E. 1972. Aping monkeys with mathematics. In *Functional and evolutionary biology of primates,* ed. R. H. Tuttle. Chicago: Aldine-Atherton.

COLE, L. C. 1954. The population consequences of life history phenomena. *Quart. Rev. Biol.* 29:103–137.

DAWKINS, R. 1976. *The selfish gene.* New York: Oxford University Press.

DEAG, J. M., AND J. H. CROOK. 1971. Social behaviour and "agonistic buffering" in the wild barbary macaque *Macaca sylvana* L. *Folia Primatol.* 15:183–200.

DEVORE, I. 1963. Mother-infant relations in free-ranging baboons. In *Maternal behavior in mammals,* ed. H. L. Rheingold. New York: John Wiley & Sons.

DEVORE, I., AND K. R. L. HALL. 1965. Baboon ecology. In *Primate behavior: field studies of monkeys and apes,* ed. I. DeVore. New York: Holt, Rinehart & Winston.

DEVORE, I., AND S. L. WASHBURN. 1963. Baboon ecology and human evolution. In *African ecology and human evolution,* ed. F. C. Howell and F. Bourliere. Viking Fund Publ. Anthropol., no. 36. Chicago: Aldine.

DITTUS, W. P. J. 1975. Population dynamics of the toque monkey, *Macaca sinica.* In *Socioecology and psychology of primates,* ed. Russell H. Tuttle. The Hague: Mouton Publishers.

————— 1977. The social regulation of population density and age-sex distribution in the Toque monkey. *Behaviour* 63:281–322.

DRAPER, P. 1974. Comparative studies of socialization. *Ann. Rev. Anthrop.* 3:263–277.

————— 1976. Social and economic constraints on child life among the !Kung. In *Kalahari hunter-gatherers,* ed. R. B. Lee and I. DeVore. Cambridge, Mass.: Harvard University Press.

DRICKAMER, L. C. 1974. A ten-year summary of reproductive data for free ranging *Macaca mulatta. Folia Primatol.* 21:61–80.

DUNBAR, R. I. M., AND E. P. DUNBAR, 1977. Dominance and reproductive success among female gelada baboons. *Nature* 266:351.

EMERSON, K., JR., B. N. SAXENA, AND E. L. POINDEXTER. 1972. Caloric cost of normal pregnancy. *Obstet. Gynecol.* 40:786–794.

EMLEN, J. M. 1970. Age specificity and ecological theory. *Ecology* 51:588–601.

FISHER, R. A. 1930. *The genetical theory of natural selection.* Oxford: Clarendon Press.

FISLER, G. F. 1967. Nonbreeding activities of three adult males in a band of free-ranging rhesus monkeys. *J. Mammal.* 48:70–78.

FOX, R. 1967. *Kinship and marriage.* Harmondsworth, Eng.: Penguin Books.

FREEDMAN, D. G. 1974. *Human infancy: an evolutionary perspective.* New York: John Wiley & Sons.

GILLMAN, J., AND C. GILBERT. 1946. The reproductive cycle of the chacma baboon (*Papio ursinus*) with special reference to the problems of menstrual irregularities as assessed by the behavior of the sex skin. *S. Afr. J. Med. Sci.* 11:1–54.

GIVONI, B., AND R. F. GOLDMAN. 1971. Predicting metabolic energy cost. *J. Appl. Physiol.* 30:429–433.

GOODALL, J. 1977. Infant killing and cannibalism in free-living chimpanzees. *Folia Primatol.* 28:259–282.

GOODMAN, L. A., AND W. H. KRUSKAL. 1954. Measures of association for cross classifications. *Am. Stat. Assoc. J.* 49:732–764.

GOSWELL, M. J., AND J. S. GARTLAN. 1965. Pregnancy, birth, and early infant behavior in the captive patas monkey, *Erythrocebus patas. Folia Primatol.* 3:189–200.

GOULD, S. J. 1975. Allometry in primates, with emphasis on scaling and the evolution of the brain. In *Approaches to primate paleobiology,* ed. G. Szalay. Basel: Karger.

GOUZOULES, H. T. 1974. Group responses to parturition in *Macaca arctoides. Primates* 15:287–292.

HALL, K. R. L. 1963. Variations in the ecology of the chacma baboon. *Symp. Zool. Soc. London* 10:1–28.

HALL, K. R. L., AND I. DEVORE. 1965. Baboon social behavior. In *Primate behavior,* ed. I. DeVore. New York: Holt, Rinehart and Winston.

HAMILTON, W. D. 1964. The genetical evolution of social behaviour: I, II. *J. Theor. Biol.* 7:1–52.

—————— 1971. Geometry for the selfish herd. *J. Theor. Biol.* 31:295–311.

HAMILTON, W. J., R. E. BUSKIRK, AND W. H. BUSKIRK. 1978. Omnivory and utilization of food resources by chacma baboons, *Papio ursinus. Am. Nat.* 112:911–924.

HARDING, R. S. O. 1973. Predation by a troop of olive baboons (*Papio anubis*). *Am. J. Phys. Anthrop.* 38:587–591.

HARLOW, H. F. 1958. The nature of love. *Proc. 66th Ann. Conv. Am. Psych. Assoc.,* Aug. 1958, pp. 673–785.

HARLOW, H. F., AND M. HARLOW, 1965. The affectional systems. In *Behavior of nonhuman primates,* ed. A. M. Schrier, H. F. Harlow, and F. Stollnitz. New York: Academic Press.

HAUSER, P. M., AND O. D. DUNCAN. 1959. *The study of population*. Chicago: University of Chicago Press.

HAUSFATER, G. 1975a. *Dominance and reproduction in baboons: a quantitative analysis*. Contributions to Primatology, vol. 7. Basel: Karger.

–––––– 1975b. Knuckle walking by a baboon. *Am. J. Phys. Anthrop.* 43:303–305.

–––––– 1976. Predatory behavior of yellow baboons. *Behaviour* 56:44–68.

HAUSFATER, G., AND W. H. BEARCE. 1976. Acacia tree exudates: their composition and use as a food source by baboons. *E. Afr. Wildl. J.* 14:241–243.

HINDE, R. A., AND S. ATKINSON. 1970. Assessing the roles of social partners in maintaining mutual proximity, as exemplified by mother-infant relations in rhesus monkeys. *Anim. Behav.* 18:169–176.

HINDE, R. A., AND T. E. ROWELL. 1962. Communication by postures and facial expressions in the rhesus monkey (*Macaca mulatta*). *Proc. Zool. Soc. London* 138:1–21.

HINDE, R. A., T. E. ROWELL, AND Y. SPENCER-BOOTH. 1964. Behaviour of socially living rhesus monkeys in their first six months. *J. Zool.* 143:609–649.

HINDE, R. A., AND M. J. A. SIMPSON. 1975. Qualities of mother-infant relationships in monkeys. In *Parent-infant interaction: Ciba Foundation Symposium 33*. Amsterdam.

HINDE, R. A., AND Y. SPENCER-BOOTH. 1967. The behaviour of socially living rhesus monkeys in their first two and a half years. *Anim. Behav.* 15:169–196.

–––––– 1971. Towards understanding individual differences in rhesus mother-infant interaction. *Anim. Behav.* 19:165–173.

HINES, M. 1942. The development and regression of reflexes, postures, and progression in the young macaque. *Contrib. Embryol. Carneg. Inst.* 196:155–209.

HOPF, S. 1967. Notes on pregnancy, delivery, and infant survival in captive squirrel monkeys. *Primates* 8:323–332.

HORWICH, R. H., AND D. MANSKI. 1975. Maternal care and infant transfer in two species of *Colobus* monkeys. *Primates* 16:49–74.

HOWELL, N. 1979. *Demography of the Dobe !Kung*. New York: Academic Press.

HRDY, S. B. 1974. Male-male competition and infanticide among the langurs (*Presbytis entellus*) of Abu, Rajasthan. *Folia Primatol.* 22:19–58.

–––––– 1976. Care and exploitation of nonhuman primate infants by conspecifics other than the mother. In *Advances in the study of behavior*, vol. 6. New York: Academic Press.

–––––– 1977. *The langurs of Abu: female and male strategies of reproduction*. Cambridge, Mass.: Harvard University Press.

HUTCHINS, M., AND D. P. BARASH. 1976. Grooming in primates: implications for its utilitarian function. *Primates* 17:145–150.

HYTTEN, F. E., AND R. I. LEITCH. 1964. *The physiology of human pregnancy*. Oxford: Blackwell.

JAIN, A. K., T. C. HSU, R. FREEDMAN, AND M. C. CHANG. 1970. Demographic aspects of lactation and postpartum amenorrhea. *Demography* 7:255–271.

JAY, P. 1963. Mother-infant relations in langurs. In *Maternal behavior in mammals,* ed. H. R. Rheingold. New York: John Wiley & Sons.

JENSEN, G. D., R. A. BOBBITT, AND B. N. GORDON. 1967. Sex differences in social interaction between infant monkeys and their mothers. *Rec. Adv. Biol. Psychiat.* 21:283–292.

JOHNSON, L. L. 1979. Kin selection in finite sibships. *J. Theor. Biol.* 77:379–381.

JOLLY, A. 1972. Hour of birth in primates and man. *Folia Primatol.* 18:108–121.

JOLLY, C. J. 1970. The seed-eaters: a new model of hominid differentiation based on a baboon analogy. *Man* 4:5–26.

KACZMARSKI, F. 1966. Bioenergetics of pregnancy and lactation in the bank vole. *Acta Theriol.* 11:409–417.

KAHNEMAN, D. 1973. *Attention and effort.* Englewood Cliffs, N.J.: Prentice-Hall.

KAUFMAN, I. C. 1974. Mother infant relations in monkeys and humans: a reply to Prof. Hinde. In *Ethology and psychiatry,* ed. N. F. White. Toronto: University of Toronto Press.

KAUFMAN, I. C., AND L. A. ROSENBLUM. 1966. A behavioral taxonomy for *Macaca nemestrina and Macaca radiata:* based on longitudinal observation of family groups in the laboratory. *Primates* 7:206–258.

KAWAI, M. 1958. On the system of social ranks in a natural troop of Japanese monkeys: I. basic rank and dependent rank. Trans. S. Takada. In *Japanese monkeys: a collection of translations,* selected by K. Imanishi, ed. S. Altmann. Published by the Editor, 1965.

KAWAMURA, S. 1958. Matriarchal social ranks in the Minoo-B troop: a study of the rank system of Japanese monkeys. Trans. K. Ozaki. In *Japanese monkeys: a collection of translations,* selected by K. Imanishi, ed. S. Altmann. Published by the Editor, 1965.

KAYE, K. 1977. *CRESCAT: software system for analysis of sequential or real-time data.* Chicago: University of Chicago Computation Center.

KEIDING, N. 1977. Statistical comments on Cohen's application of a simple stochastic population model to natural primate troops. *Am. Nat.* 111:1211–1219.

KLEIBER, M. 1961. *The fire of life.* New York: John Wiley & Sons.

KLEIMAN, D. G. 1977. Monogamy in mammals. *Quart. Rev. Biol.* 52:39–69.

KOHN, M. L. 1977. *Class and conformity: a study in values.* Chicago: University of Chicago Press.

KONNER, M. J. 1976. Maternal care, infant behavior, and development among the !Kung. In *Kalahari hunter-gatherers: studies of the !Kung San and their neighbors,* ed. R. B. Lee and I. DeVore. Cambridge, Mass.: Harvard University Press.

KRAEMER, H. C. 1979. One-zero sampling in the study of primate behavior. *Primates* 20:237–244.

KREBS, J. R. 1978. Optimal foraging: decision rules for predators. In *Beha-*

vioural ecology: an evolutionary approach, ed. J. R. Krebs and N. B. Davies. Sunderland, Mass.: Sinauer Associates.

KRIEWALDT, F. H., AND A. G. HENDRICKX. 1968. Reproductive parameters of the baboon. *Lab. Anim. Care* 18:361–370.

KUMMER, H. 1968. *Social organization of hamadryas baboons.* Chicago: University of Chicago Press.

KUMMER, H., AND F. KURT. 1965. A comparison of social behavior in captive and wild hamadryas baboons. In *The baboon in medical research,* ed. H. Vagtborg. Austin: University of Texas Press.

KURLAND, J. A. 1977. *Kin selection in the Japanese monkey.* Contributions to Primatology, vol. 12. Basel: Karger.

LANCASTER, J. B. 1971. Play-mothering: the relations between juvenile females and young infants among free-ranging vervet monkeys (*Cercopithecus aethiops*). *Folia Primatol.* 15:161–182.

LANCASTER, J. B., AND R. B. LEE. 1965. The annual reproductive cycle in monkeys and apes. In *Primate behavior: field studies of monkeys and apes,* ed. I. DeVore. New York: Holt, Rinehart and Winston.

LEE, R. B. 1978. Lactation, ovulation, infanticide, and women's work: a study of hunter-gatherer population regulation. Paper presented at the Hudson Symposium on Biosocial Mechanisms of Population Regulation, State University of New York College at Plattsburgh, April 1978.

LEE, R. B., AND I. DEVORE, eds. 1976. *Kalahari hunter-gatherers: studies of the !Kung San and their neighbors.* Cambridge, Mass.: Harvard University Press.

LEVINE, R. A. 1973. *Culture, behavior, and personality.* Chicago: Aldine.

LEVINS, R. 1968. *Evolution in changing environments: some theoretical explorations.* Princeton: Princeton University Press.

LEVITT, P. R. 1975. General kin selection models for genetic evolution of sib altruism in diploid and haplodiploid species. *Proc. Nat. Acad. Sci. U.S.A.* 72:4531–4535.

LEWONTIN, R. C. 1965. Selection for colonizing ability. In *The genetics of colonizing species,* ed. H. G. Baker and G. L. Stebbins. New York: Academic Press.

LINDBURG, D. G. 1971. The rhesus monkey in north India: an ecological and behavioral study. In *Primate behavior: developments in field and laboratory research,* vol. 2, ed. L. A. Rosenblum. New York: Academic Press.

LOVE, J. A. 1978. A note on the birth of a baboon (*Papio anubis*). *Folia Primatol.* 29:303–306.

McCUSKEY, S. A. 1975. Demography and behavior of one-male groups of yellow baboons (*Papio cynocephalus*). M.S. thesis, University of Virginia.

MacNAIR, M. R., AND G. A. PARKER. 1978. Models of parent-offspring conflict: II. Promiscuity. *Anim. Behav.* 26:111–122.

McNAUGHTON, S. J. AND L. L. WOLF. 1973. *General ecology.* New York: Holt, Rinehart and Winston.

MARAIS, E. 1969. *The soul of the ape.* New York: Atheneum.

MASON, W. A. 1978. Social experience and primate cognitive development. In

The development of behavior: comparative and evolutionary aspects, ed.
G. M. Burghardt and M. Bekoff. New York: Garland STPM Press.

MASON, W. A., AND G. BERKSON. 1975. Effects of maternal mobility on the development of rocking and other behaviors in rhesus monkeys: a study with
artificial mothers. *Develop. Psychobiol.* 8:197–211.

MASON, W. A., E. W. MENZEL, AND R. K. DAVENPORT. 1968. Early experience
and the social development of rhesus monkeys and chimpanzees. In *Early
experience and behavior,* ed. G. Newton and S. Levine. New York: Char-
les C Thomas.

MAYNARD SMITH, J., AND G. A. PARKER. 1976. The logic of asymmetric contests.
Anim. Behav. 24:159–175.

MAYR, E. 1974. Behavior programs and evolutionary strategies. *Am. Sci.*
62:650–659.

MERTZ, D. B. 1971. Life history phenomena in increasing and decreasing populations. In *Statistical ecology,* vol. 2: *Sampling and modeling biological
populations and population dynamics,* Reprinted with corrections. University Park: Pennsylvania State University Press.

MISSAKIAN, E. A. 1972. Genealogical and cross-genealogical dominance relations in a group of free-ranging rhesus monkeys (*Macaca mulatta*) on Cayo
Santiago. *Primates* 13:169–180.

MITCHELL, G. D. 1968. Attachment differences in male and female infant monkeys. *Child Develop.* 39:611–620.

MYERS, J. H. 1978. Sex ratio adjustment under food stress: maximization of
quality or numbers of offspring? *Am. Nat.* 112:381–388.

NAISMITH, D. J., AND C. D. RITCHIE. 1975. The effect of breast-feeding and artificial feeding on body-weights, skinfold measurements, and food intakes of
forty-two primiparous women. *Proc. Nutr. Soc.* 34:116A–117A.

NASH, L. T. 1974. Parturition in a feral baboon. *Primates* 15:279–286.

———— 1978. The development of the mother-infant relationship in wild baboons (*Papio anubis*). *Anim. Behav.* 26:746–759.

NIE, N. H., C. H. HULL, J. G. JENKINS, K. STEINBRENNER, AND D. H. BENT. 1975.
SPSS: statistical package for the social sciences. 2nd ed. New York:
McGraw-Hill Book Co.

NORIKOSHI, K. 1974. The development of peer-mate relationships in Japanese
macaque infants. *Primates* 15:39–46.

OWENS, N. W. 1975. Social play behaviour in free-living baboons, *Papio
anubis. Anim. Behav.* 23:387–408.

PACKER, C. 1979a. Inter-troop transfer and inbreeding avoidance in *Papio
anubis. Anim. Behav.* 27:1–37.

———— 1979b. Male dominance and reproductive activity in *Papio anubis.
Anim. Behav.* 27:37–46.

PARKER, G. A., AND M. R. MACNAIR. 1978. Models of parent-offspring conflict:
I. Monogamy. *Anim. Behav.* 26:97–110.

POIRIER, F. E. 1968. The Nilgiri langur (*Presbytis johnii*) mother-infant dyad. *Primates* 9:45–68.

POST, D. 1978. Feeding and ranging behavior of the yellow baboon (*Papio cynocephalus*). Ph.D. thesis, Yale University.

POST, W., AND J. BAULU. 1978. Time budgets of *Macaca mulatta*. *Primates* 19:125–140.

PYKE, G. H., H. R. PULLIAM, AND E. L. CHARNOV. 1977. Optimal foraging: a selective review of theory and tests. *Quart Rev. Biol.* 52:137–154.

RANSOM, T. W., AND B. S. RANSOM. 1971. Adult male-infant relations among baboons (*Papio anubis*). *Folia Primatol.* 16:179–195.

RANSOM, T. W., AND T. E. ROWELL. 1972. Early social development of feral baboons. In *Primate socialization,* ed. F. E. Poirier. New York: Random House.

RASMUSSEN, D. R., AND K. L. RASMUSSEN. 1979. Social ecology of adult males in a confined troop of Japanese macaques (*Macaca fuscata*). *Anim. Behav.* 27:434–445.

REDICAN, W. K. 1976. Adult male-infant interactions in nonhuman primates. In *The role of the father in child development,* ed. Michael E. Lamb. New York: John Wiley & Sons.

REYNOLDS, M. 1967. Mammary respiration in lactating goats. *Am. J. Physiol.* 212:707–710.

RHEINGOLD, H. L., AND C. D. ECKERMAN. 1970. The infant separates himself from his mother. *Science* 168:78–84.

ROSE, M. D. 1977. Positional behaviour of olive baboons (*Papio anubis*) and its relationship to maintenance and social activities. *Primates* 18:59–116.

ROSENBLUM, L. A. 1974. Sex differences, environmental complexity, and mother-infant relations. *Arch. Sex. Behav.* 3:117–128.

ROWELL, T. E. 1966a. Hierarchy in the organization of a captive baboon groop. *Anim. Behav.* 14:430–443.

——— 1966b. Forest living baboons in Uganda. *J. Zool.* 149:344–364.

——— 1968. The effect of temporary separation from their group on the mother infant relationship of baboons. *Folia Primatol.* 9:114–122.

——— 1969. Intra-sexual behaviour and female reproductive cycles of baboons (*Papio anubis*). *Anim. Behav.* 17:159–167.

ROWELL, T. E., N. A. DIN, AND A. OMAR. 1968. The social development of baboons in their first three months. *J. Zool. London* 155:461–483.

ROWELL, T. E., AND R. A. HINDE. 1962. Vocal communication by the rhesus monkey (*Macaca mulatta*). *Proc. Zool. Soc. London* 138:279–294.

SAAYMAN, G. S. 1971. Grooming behavior in a troop of free-ranging chacma baboons (*Papio ursinus*). *Folia Primatol.* 16:161–178.

SADE, D. S. 1965. Some aspects of parent-offspring and sibling relations in a group of rhesus monkeys with a discussion of grooming. *Am. J. Phys. Anthrop.* 23:1–18.

——— 1967. Determinants of dominance in a group of free-ranging rhesus monkeys. In *Social communication among primates,* ed. S. A. Altmann. Chicago: University of Chicago Press.

SADE, D. S., K. CUSHING, P. CUSHING, J. DUNAIF, A. FIGUEROA, J. KAPLAN, C.

LAUER, D. RHODES, AND J. SCHNEIDER. 1977. Population dynamics in relation to social structure on Cayo Santiago. *Yrbk. Phys. Anthrop.* 20 (1976):253–262.

SAXENA, P. C. 1977. Breast-feeding: its effects on post-partum amenorrhea. *Social Biol.* 24:45–51.

SCHMIDT-NIELSEN, K. 1977. Problems of scaling: locomotion and physiological correlates. In *Scale effects in animal locomotion,* ed. T. J. Pedley. London: Academic Press.

SEYFARTH, R. M. 1976. Social relationships among adult female baboons. *Anim. Behav.* 24:917–938.

_____ 1977. A model of social grooming among adult female monkeys. *J. Theor. Biol.* 65:671–698.

_____ 1978. Social relationships among adult male and female baboons. II. Behaviour throughout the female reproductive cycle. *Behaviour* 64:227–247.

SIMPSON, M. J. A., AND A. E. SIMPSON. 1977. One-zero and scan methods for sampling behaviour. *Anim. Behav.* 25:726–731.

SLATKIN, M. 1975. A report on the feeding behavior of two East African baboon species. In *Contemporary primatology,* ed. S. Kondo. Basel: Karger.

SLATKIN, M., AND G. HAUSFATER. 1976. A note on the activities of a solitary male baboon. *Primates* 17:311–322.

SNOW, C. C. 1967. The physical growth and development of the open-land baboon, *Papio doguera.* Ph.D. thesis, University of Arizona.

SOULE, R. G., K. B. PANDOLF, AND R. F. GOLDMAN. 1978. Energy expenditure of heavy load carriage. *Ergonomics* 21:373–381.

STAMPS, J. A., R. A. METCALF, AND V. V. KRISHNAN. 1978. A genetic analysis of parent-offspring conflict. *Behav. Ecol. Sociobiol.* 3:369–392.

STOLZ, L. P., AND G. S. SAAYMAN. 1970. Ecology and behaviour of baboons in the Northern Transvaal. *Ann. Transvaal Mus.* 26:99–143.

STRUHSAKER, T. T. 1967. Ecology of vervet monkeys (*Cercopithecus aethiops*) in the Masai-Amboseli Game Reserve, Kenya. *Ecology* 48:891–904.

_____ 1971. Social behaviour of mother and infant vervet monkeys (*Cercopithecus aethiops*). *Anim. Behav.* 19:233–250.

_____ 1973. A recensus of vervet monkeys in the Masai-Amboseli Game Reserve, Kenya. *Ecology* 54:930–932.

_____ 1976. A further decline in numbers of Amboseli vervet monkeys. *Biotropica* 8:211–214.

STRUM, S. C. 1975. Primate predation: interim report on the development of a tradition in a troop of olive baboons. *Science* 187:755–757.

SUGIYAMA, Y. 1965. Behavioral development and social structure in two troops of hanuman langurs (*Presbytis entellus*). *Primates* 6:213–248.

_____ 1967. Social organization of hanuman langurs. In *Social communication among primates,* ed. S. A. Altmann. Chicago: University of Chicago Press.

TAYLOR, C. R., K. SCHMIDT-NIELSEN, AND J. L. RAAB. 1970. Scaling of energetic cost of running to body size in mammals. *Am. J. Physiol.* 219:1104–1107.

TEITELBAUM, M. S., N. MANTEL, AND C. STARK. 1971. Limited dependence of the human sex ratio on birth order and parental ages. *Am. J. Hum. Gen.* 23:271–280.

TRIVERS, R. L. 1972. Parental investment and sexual selection. In *Sexual selection and the descent of man, 1871–1971,* ed. B. Campbell. Chicago: Aldine.

———— 1974. Parent-offspring conflict. *Am. Zool.* 14:249–264.

TRIVERS, R. L., AND D. E. WILLARD. 1973. Natural selection of parental ability to vary the sex ratio of offspring. *Science* 179:90–91.

VAN LAWICK-GOODALL, J. 1967. Mother-offspring relationships in free-ranging chimpanzees. In *Primate ethology,* ed. D. Morris. Chicago: Aldine.

———— 1971. Some aspects of mother-infant relationships in a group of wild chimpanzees. In *The origins of human social relations,* ed. H. R. Schaffer. London: Academic Press.

VAN WAGENEN, G., AND H. R. CATCHPOLE. 1956. Physical growth of the rhesus monkey (*Macaca mulatta*). *Am. J. Phys. Anthrop.* 14:245–273.

WADE, M. J. 1978. Kin selection: a classical approach and a general solution. *Proc. Nat. Acad. Sci. U.S.A.* 75:6154–6158.

WALTERS, J. In press. Interventions and the development of dominance relationships in female baboons. *Folia Primatol.*

WEST EBERHARD, M. J. 1975. The evolution of social behavior by kin selection. *Quart. Rev. Biol.* 50:1–33.

WESTERN, D. 1973. The structure, dynamics and changes of the Amboseli ecosystem. Ph.D. thesis, University of Nairobi.

———— 1977. *The ecology of tourists. Anim. Kingdom* 80:23–30.

WESTERN, D., AND C. VAN PRAET. 1973. Cyclical changes in the habitat and climate of an east African ecosystem. *Nature* 241:104–106.

WHICHELOW, M. J. 1976. Success and failure of breast-feeding in relation to energy intake. *Proc. Metr. Soc.* 35:62A.

WHITING, B. B., AND J. W. M. WHITING. 1975. *Children of six cultures: a psycho-cultural analysis.* Cambridge Mass.: Harvard University Press.

WHITING, J. W. M., AND I. L. CHILD. 1953. *Child training and personality: a cross-cultural study.* New Haven: Yale University Press.

WILSON, E. O. 1975. *Sociobiology: the new synthesis.* Cambridge, Mass.: Belknap Press of Harvard University Press.

WOOLDRIDGE, F. L. 1971. *Colobus guereza:* birth and infant development in captivity. *Anim. Behav.* 19:481–485.

YOUNG, G. H., AND C. A. BRAMBLETT. 1977. Gender and environment as determinants of behavior in infant common baboons (*Papio cynocephalus*). *Arch. Sex. Behav.* 6:365–385.

YOUNG, G. H., AND R. J. HANKINS. 1979. Infant behaviors in mother-reared and harem-reared baboons (*Papio cynocephalus*). *Primates* 20:87-93.

Index